Cosmetic Nanotechnology

ACS SYMPOSIUM SERIES **961**

Cosmetic Nanotechnology

Polymers and Colloids in Cosmetics

Sarah E. Morgan, Editor
University of Southern Mississippi

Kathleen O. Havelka, Editor
Lubrizol Corporation

Robert Y. Lochhead, Editor
University of Southern Mississippi

Sponsored by the ACS Division of Polymer Chemistry, Inc.

American Chemical Society, Washington, DC

Library of Congress Cataloging-in-Publication Data

Cosmetics nanotechnology : polymers and colloids in cosmetics / Sarah E. Morgan, editor, Kathleen O. Havelka, editor, Robert Y. Lochhead, editor.

p. cm.—(ACS symposium series ; 961)

Includes bibliographical references and index.

ISBN 13: 978–0–8412–3996–8 (alk. paper)

ISBN 10: 0–8412–3996–7 (alk. paper)

1. Cosmetic delivery systems—Congresses. 2. Nanotechnology—Congresses. 3. Polymers—Industrial applications—Congresses. 4. Colloids—Industrial applications—Congresses.

I. Morgan, Sarah E., 1958- II. Havelka, Kathleen O. III. Lochhead, Robert Y., 1943-

TP983.3.C675 2007
668′.55—dc22

2006052604

The paper used in this publication meets the minimum requirements of American National Standard for Information Sciences—Permanence of Paper for Printed Library Materials, ANSI Z39.48–1984.

Distributed by Oxford University Press

PRINTED IN THE UNITED STATES OF AMERICA

Foreword

The ACS Symposium Series was first published in 1974 to provide a mechanism for publishing symposia quickly in book form. The purpose of the series is to publish timely, comprehensive books developed from ACS sponsored symposia based on current scientific research. Occasionally, books are developed from symposia sponsored by other organizations when the topic is of keen interest to the chemistry audience.

Before agreeing to publish a book, the proposed table of contents is reviewed for appropriate and comprehensive coverage and for interest to the audience. Some papers may be excluded to better focus the book; others may be added to provide comprehensiveness. When appropriate, overview or introductory chapters are added. Drafts of chapters are peer-reviewed prior to final acceptance or rejection, and manuscripts are prepared in camera-ready format.

As a rule, only original research papers and original review papers are included in the volumes. Verbatim reproductions of previously published papers are not accepted.

ACS Books Department

Contents

Overview

Polymer Architecture and Film Performance

Advances in Eye Care

Natural Products in Personal Care and Biomedical Applications

Silicone Polymers in Cosmetics

Commercial and Regulatory Considerations for Cosmetic Nanotechnology

Indexes

Overview

Chapter 1

The Role of Polymers in Cosmetics: Recent Trends

Robert Y. Lochhead

The Institute for Formulation Science, School of Polymers and High Performance Materials, The University of Southern Mississippi, 118 College Drive, Hattiesburg, MS 39406–0076

Polymers represent the second largest class of ingredients in cosmetics and personal care products. A diverse range of polymers are applied in this segment as film formers, fixatives, rheology modifiers, associative thickeners, emulsifiers, stimuli-responsive agents, conditioners, foam stabilizers and destabilizers, skin-feel beneficial agents, and antimicrobials. This chapter reviews recent advances in the use of polymers in personal care formulation. The drive to meet the requirements of the Clean Air Act have drawn polymeric materials into formulations that are low in volatile-organic compounds. This has resulted in aqueous-based thickeners that also form films and act as fixatives. The onset of free-radial living polymerization offers the prospect of custom-designing the morphology of products to meet the desired attributes. Complex coacervate mechanisms dominate the functioning of conditioning shampoos. Stimuli-responsive polymers are being directed towards applications from make-up that camouflages wrinkles to the facile processing of multiple emulsions, to thermally-responsive systems that respond to the surface of skin.

The use of polymers in cosmetics is highly developed, and innovative advances in polymer science and nanoscience are driving the creation of scientifically sophisticated products.

Polymeric materials represent a major class of ingredients that are used in cosmetics. Polymers are used as:

- film formers in hair fixatives, mascara, nail enamels and transfer-resistant color cosmetics
- thickeners and rheology modifiers and emulsion stabilizers, gels, binders, hair columns and hair relaxers
- emulsifiers in lotions, sunscreens and hair-colorants
- conditions for both hair and skin
- moisturizers for skin
- emollients to improve the 'rub-in' characteristics of lotions
- dispersers for pigments
- waterproofers in sunscreens and make-up
- micropore sponges for controlled release and oil absorption

Cosmetic raw materials are named according to the International Nomenclature Cosmetic Ingredient (INCI) convention.[1] (1). Throughout this article, we will adopt INCI nomenclature. According to the INCI convention, polymeric materials are named according to the name in common usage if it is well-known, or by the structure if well-defined. If no common name exists and the structure is not well-defined, the polymers are named according to:

1. homopolymers are named by placing the term "poly" before the constituent monomer
2. copolymers are named by listing the monomers separated by a slash (/) followed by the word " Copolymer"
3. copolymers consisting of four or more monomers may be given an INCI name according to their class followed by an arbitrary number, e.g. Polyester-1
4. crosslinked polymers consisting of two or more constituent monomers are named by listing the monomers in alphabetical order separated by a slash, followed by the word "Crosspolymer"
5. the term "Acrylates" describes non-crosslinked copolymers that contain combinations of acrylic acid, methacrylic acid, and their simple esters
6. the term "Crotonates" non-crosslinked copolymers that contain combinations of crotonic acid and its simple esters
7. the term "Aminoacrylates" refers to simple aminoacrylates in which the substituent alkyl groups attached to amino nitrogen range from C 1-4, and acrylates conforms to the definition above
8. the name "Carbomer" describes high molecular weight crosslinked polymers of acrylic acid.
9. Vinylpyrrolidone is abbreviated to "VP" and polyvinylpyrrolidone is abbreviated to "PVP"
10. The copolymer of vinyl methyl ether and maleic anhydride is abbreviated to PVM/MA copolymer.

11. The name " Poloxamer" describes ABA tnblock copolymers of polyoxyethylene and polyoxypropylene in which the A block is polyoxyethylene and the B block is polyoxypropylene.

Film Formers

Hair Fixatives: Hour fixatives are essentially film-forming polymers that are used to adhere hair into a preferred style. The desired attributes of hair fixatives include hair style hold improvement, ease of application on wet hair, ease of combing, no sticky feel, quick drying, not powdery or flaky during grooming, assurance of hair body and bounce, increased hair volume, no clumping of the hair, formation of a non-hygroscopic binding film that is removable by shampoo, good hair gloss, and no excessive stiffness.

Hair style fixation should be achieved with a tiny amount of fixative polymer applied from an aerosol spray, a pump spray, a gel or a mousse. The solvent must be compatible with water and the polymer must not phase-separate as water permeates the system from wet hair. The fixative system must also be safe, especially with respect to ocular exposure or inhalation. In order to prevent hair damage during grooming, the cohesion of the polymer film must be less than the adhesion of the film to the hair. Sticky feel can be avoided by formulating for quick drying, which can be achieved either by choice of solvent and propellant, or by causing the system to gel. The desired mechanical property aesthetics of hairstyle are body, bounce, increased hair volume, non-clumping and lack of extreme stiffness. These are achieved by allowing the fixative polymer to lightly "spot-weld' the hair matrix being lightly crosslinked rather than coating each hair with the applied polymer film. The film must be non-hygroscopic; otherwise it will be plasticized by absorbed water vapor under humid conditions and hold will be lost.

Volatile organic compounds: For the last quarter of a century, the regulatory requirements of the Clean Air Act has motivated the U.S. aerosol hair spray industry to develop systems with low volatile organic compound (VOC) content. In most cases this has meant replacing a significant portion of the ethanol solvent with water. The addition of water to conventional hairspray systems creates two main challenges that must be overcome. First, the hair surface is relatively hydrophohic and the kinetics and thermodynamics of hair fiber wetting are less favorable for "aqueous" systems than systems based upon ethanol, which wets the fiber more effectively than water. Hairstyle fixation relies upon capillary forces to bring the hair together, normal to the longitudinal fiber axis. This can only occur if the hair is wetted by the applied liquid. Addition of substantial amounts of water to the formulation causes a reduction of capillarity and results in loss of holding power due to the inter-fiber effects discussed above. The second challenge arises from the fact that the water rapidly penetrates into the hair cortex, plasticizing the hair and causing "droop" of the style.

Table 1: Examples of cosmetics film-formers.

Nonionic Film Formers
PVP [Poly(vinyl pyrolidone)], PVP/VA Copolymer
Polyvinylcaprolactam
Poly(vinyl alcohol), Poly (vinyl alcohol-*co*-vinylacetate)
Polyurethane-2,Polyurethane-5
PPG-17/IPDI/DMPA Copolymer
PVP/Dimethiconylacrylate/Polycarbamyl Polyglycol Ester
PEG-200 Hydrogenated Castor Oil/IPDI Copolymer
Polyglyceryl-2 Diisostearate/IPDI Copolymer
Vinylcaprolactam/PVP/Dimethylaminoethylmethacrylate Copolymer
Poly-(vinylpyrrolidone-vinylcaprolactam-dimethylaminoethylmethacrylate)
VP/Vinyl Caprolactam/DMAPA Acrylates Copolymer
Cellulose Acetate, Cellulose Acetate Butyrate, Cellulose Acetate Propionate
Nitrocellulose
Anionic Film Formers
Shellac
Acrylates Copolymer
Acrylates/C1-2Succinates/Hydroxyacrylates Copolymer
Acrylic Esters (and) Methacrylic Esters Copolymer
AMP-Acrylates/Allyl Methacrylate Copolymer
Diglycol/Cyclohexanedimethanol/Isophtalates/Sulfoisophtalates Copolymer
Diglycol/Isophtalates/Sulfoisophtalates Copolymer
Isobutylene Ethylmaleimide/Hydroxyethylmaleimide Copolymer
Glycerin and Diglycol/Cyclohexanedimethanol/Isophtalates/Sulfoisophtalates Copolymer
Methacrylate Acid/Sodium Acrylamidomethyl Propane Sulfonate Copolymer
Octylacrylamide/Acrylates/Butylaminoethylmethacrylate Copolymer
Polyurethane (and) Acrylates Copolymer
PVM/MA Copolymer
PVP/Ethyl Methacrylate/Methacrylic Acid Terpolymer
PVP/Polycarbamyl Polyglycol Ester
VA/Crotonates Copolymer
VA/Crotonates/Vinyl Neodecanoate Copolymer
VP/Acrylates/Lauryl Methacrylate Copolymer
Amphoteric Film Formers
Acrylates/Lauryl Acrylate/Stearyl Acrylate/Ethylamine Oxide Methacrylate
Methacryloyl Ethyl Betaine/Acrylates Copolymer

Random copolymers have solubility parameters that are intermediate between their constituent homopolymers. Conventional hair spray resins were designed to be soluble in a propellant mixture of ethanol solvent and propane/isobutane. Addition of water to the polymer solution could cause the solvent mixture's solubility parameter to become sufficiently different to that of the polymer that the formulation would phase separate. Polymer suppliers initially responded to the need for these types of fixative polymers by merely lowering the molecular weights of their polymer products to render them more combatible with aqueous mixtures. This strategy was based upon the fact that the configurational entropy of mixing increases as the polymer molecular weight decreases and, as a consequence, lower molecular-weight polymers are soluble in a wider range of solvents than higher molecular-weight polymers of the same chemical composition and structure. However, the lowering of molecular weight adversely affects film properties and the holding properties were reduced. This imposed limitations on the extent to which this approach could be adopted and other routes were sought.

Block and graft copolymers in low VOC fixatives: Compared to random copolymers, block and graft copolymers can show dual solubility zones, with each zone corresponding to the solubility parameter of the respective homopolymers. Non- interacting homopolymer mixtures do not, in general, form compatible blends because the large molecules have low entropies of mixing, and in the absence of large enthalpies of mixing, the free energies of mixing favor segregation of the polymer mixture into two phases – one rich in one of the homopolymers and the other phase that is rich in the other homopolymer.

Each of the component "molecular blocks" of block and graft copolymers also display such segregative phase separation. However, in this case because the molecular blocks are covalently linked, the segregation is limited to microdomain or nanodomain dimensions. This limited segregation leads to polymer alloys in which glassy microdomains confer mechanical rigidity, and rubbery microdomains confer shock-absorbing properties. These materials are thermoplastic elastomers that can display enhanced mechanical properties when compared to the homopolymers. Amphipathic block copolymers contain polar blocks that confer the desirable attribute of facile removal of the polymer from hair by shampooing. The cosmetic and styling properties can be enhanced by the inclusion of a non-thickening amphiphilic block copolymer in a fixative formulation based on a conventional hair fixative polymer.[2] The enhancements disclosed are improvements in conditioning or styling power without increasing the total polymer concentration in the formula, or the same conditioning and styling benefits at reduced polymer concentration. The non-thickening polymer in this case is defined as a copolymer which when dispersed or dissolved in water leads to dynamic viscosities that are less than 0.1 Pa.s measured at a shear rate of 200 sec^{-1}. The enhancement of properties derives from the polymer morphology:

- Block copolymers tend to exhibit morphology of separated microdomains, each of which is rich in one of the blocks.
- Separation of discontinuous glassy domains in rubbery matrix would confer the properties of an elastomer
- Separation of rubbery domains in a glassy matrix would confer impact resistance.
- Separation of ionic microdomains within a nonpolar matrix would confer increases in toughness and tensile modulus.

- Ionic microdomains might be expected to confer susceptibility to shampoo that would lead to ease of removability from the hair after use.
- Alternatively, preferred adsorption of the block copolymer at the hair/fixative interface could lead to improved adhesion and consequently higher fixing of the hair.

The polydimethylsiloxone (dimethicone) chain is extremely flexible and hydrophobic and, as a result, this silicone polymer is a lubricant that confers conditioning benefits such as softness and detangling to hair. A terpolymer consisting of *t*-butyl acrylate, acrylic acid and a silicone macromer is taught to provide softness, disentangling and feel benefits with simultaneous excellent styling and fixing properties.[3]

Polyurethanes are condensation copolymers that can be synthesized to form elastomers with polar and non-polar microdomains. The microdomain morphology and resulting material rheology can be enhanced by including a silicone block to produce polyurethane/silicone copolymers that provides smoothness and sleekness to the hair while being easy to wash out.[4] In another approach directed towards these combined attributes, polyureas have been introduced. These are synthesized from a polyethoxylated diamine, an amine-containing or hydroxyl-containing polysiloxane and a di-isocyanate.[5]

Chain-growth block copolymers have conventionally been synthesized by ionic 'living' polymerization. Ionic polymerization is subject to severe constraints; for example anionic polymerization is "killed' by the presence of a small concentration of protic impunities. For this reason, the monomer streams are required to have exceptionally high purity. Moreover, only a narrow range of monomers can be synthesized by this technique. These limitations are being overcome by the a introduction of living "free-radical" polymerization which offers the prospect of economically producing a wide range of block copolymers in a broad array of reactant media. The main techniques for the polymerization of block copolymer elastomers by free radical living polymerization are:

- The technique of free-radical polymerization by reaction with a nitroxide, consists in blocking the growing free-radical species in the form of a bond of $C\text{-}ONR_1R_2$ type.[6,7,8,9]
- The RAFT (reversible addition-fragmentation chain transfer) polymerization technique consists in blocking the growing free-radical species in the form of a bond of C--S type.[10,11,12] Dithio compounds such as thiobenzoates, dithiocarbamates or xanthan disulfides are used to do this.

Free-radical living polymerization has been applied to synthesize hair fixative polymers comprising at least one rigid block having a glass transition temperature (Tg) of greater than or equal to 20°C, and at least one flexible block having a glass transition temperature (Tg) of less than 20°C.[13] These block copolymers exhibit mechanical properties that are characteristic of elastomers;

for example they produce films having instantaneous recovery from strains of 5% to 100 % .

Thickeners

Hair Styling Gels are aqueous compositions that are usually thickened with crosslinked poly(acrylic acid); namely the Carbomers. The structure of Carbomers consists of microgels of cross-linked poly(acrylic acid). Carbomers confer yield stress and shear thinning characteristics (*i.e.* Ellis rheology) on aqueous systems. However, most conventional hair fixative polymers are incompatible with Carbomer gels and this has restricted the choice of film-formers to PVP and PVP/VA. Such PVP polymers are hygroscopic and are susceptible to plasticization in humid environments that causes the hairstyle to 'droop'. PVP and PVP/VA copolymers tend to make the hair feel tacky or sticky. Consequently, there have been attempts to develop new polymers to be compatible with the Carbomer thickeners. In order to develop water- soluble fixatives nonionic derivatized starches,[14] poly-N-vinylacetamide, amphoteric urethanes,[15] and dehydroxanthan gum have been introduced.[19] Copolymers containing up to 10 weight percent of 2-acrylamido-2-methyl-1-propane sulfonic acid or its salts in combination with anionic or nonionic monomers provide a better balance between conflicting requirements of good curl retention at high humidity but they can be removed from hair by merely rinsing with water. An extension of this class of polymers has been introduced in the form of fluorine-modified comb polymers based on acryloyldimethyltaurine acid.[16]

Thickeners for skin emulsions: Since the 1970's, conventional skin lotion emulsions have been stabilized against creaming and sedimentation by polymeric thickeners such as Carbomers and this class of emulsion stabilizers continue to dominate today's skin lotion compositions. Conventional Carbomers require care and time to be dispersed homogeneously in aqueous solutions. There is a need for faster dispersing polymer and some polymer manufacturers have addressed this need by supplying inverse emulsion polymers or dispersions of thickening polymers in oil in silicone fluids.[17] In common with other inverse emulsion polymers these are supplied with high HLB "activating" surfactants to enable the polymer to be dispersed and to thicken quickly when the phase is inverted upon flooding with large amounts of water.

During the 1990's hydrophobically-modified versions of Carbomers (Acrylates/C10-30 Alkyl Acrylate Crosspolymer) were introduced with both emulsifying and rheology-modifying properties.[18,19] These polymeric emulsifiers offered a route to 'surfactant-free' emulsions with triggered oil release upon application to the skin. These new emulsifiers also overcame the tendency of conventionally emulsified products to re-emulsify after the emulsion broke. This led to products such as sports sunscreen that resisted sweat.

Table 2: Examples of thickeners that are used in cosmetics

Nonionic Thickeners
Hydroxyethylcellulose, Hydroxymethylcellulose
Methylhydroxyethylcellulose, Hydroxypropylcellulose
Hydroxypropylmethylcellulose
Cetyl Hydroxyethylcellulose
Corn Starch, Hydroxypropyl Starch Phosphate
Distarch Phosphate, Distarch dimethylene urea
Aluminum Starch Octenyl Succinate
Guar gum, Hydroxypropyl Guar
Locust Bean Gum
Rosin, Sclerotium Gum
Maltodextrin,Dextran
Poly(acrylamide)
PEG-150 Distearate, PEG-150/DecylAlcohol/SMDI Copolymer
PEG-150/StearylAlcohol/SMDICopolymer,
PEG-180/Laureth-50/TMMG Copolymer, Polyether-1
Anionic Thickeners
Acrylic Acid/Acrylamidomethyl Propane Sulfonic Acid Copolymer
Acrylates/C10-30 Alkyl Acrylate Crosspolymer
Acrylates/Beheneth-25 Methacrylate Copolymer
Acrylates/Steareth-20 Methacrylate Copolymer
Acrylates Copolymer (and) Steareth-20
Acrylates/VA Crosspolymer
Acrylic Acid/Acrylonitrogens Copolymer
Ammonium Acryloyldimethyltaurate/Beheneth-25 Methacrylate Copolymer
Ammonium Acryloyldimethyltaurate/VP Copolymer
Caprylic/Capric Triglyceride (and) Sodium Acrylates Copolymer
Carbomer
PVM/MA Decadiene Crosspolymer
Alginic acid
Carageenan gum (Kappa, Iota, Lambda)
Cellulose Gum (Carboxymethylcellulose)
Gum Arabic/ Gum Acacia
Xanthan gum

Copolymers of ethyl acrylate, methacrylic acid and ethoxylated long chain alkyl acrylates were also introduced as hydrophobically modified alkali swellable thickeners.[20] Hydrophobically-modified derivatives of Acryloydimethyltaurine (AMPS) copolymers were introduced as polymeric emulsifiers[21] that could emulsify and thicken reliably at the natural pH (< 6) of the acid mantle of skin. Hydrophobically-modified acryloyldimethyltaurine copolymers with fluorocarbon hydrophobes have been patented.[22] The advantage of these fluoroalkyl derivatives derives from the observation that homologs of fluorocarbons and hydrocarbons above chain lengths of about 6 to 8 carbons are usually incompatible because the cohesive energy density of fluorocarbons is sufficiently different from hydrocarbons and the entropy of mixing is insufficient to give negative free energy of mixing. Therefore, it is reasonable to assume that the hydrophobic 'bonds' formed between the fluoroalkyl hydrophobes of the new copolymers will be less susceptible to disruption by hydrocarbon moieties. This should translate to more control by the formulator over polymeric emulsifier compositions.

Polymeric emulsifiers are useful for the preparation of high internal phase oil-in-water emulsions.[23, 24, 25] These emulsions can contain in excess of 90 per cent of dispersed oil phase. At such an oil loading, the emulsions resemble foam with closely packed oil 'droplets' instead of air bubbles. This technology is especially useful for the preparation of stable concentrated emulsions that can be simply diluted with water to render them into a final product. Another advantage is their use for moisturizing lotions that can be applied in the shower to spread and cover the skin with an occlusive oil-layer for prolonged moisturization. A further advantage over conventional emulsifiers is the ability of these polymeric emulsifiers to stably emulsify polar oils and the use of this technology in perfume emulsions has been disclosed.[26] Surprisingly, it has been disclosed that these high internal phase emulsions can be prepared using any water-soluble polymer having a molecular weight in the range 10,000 to 130,000 Daltons and the necessary level of surface activity. Film forming polymers may increase the substantivity of the emulsions on the skin to retain the oil phase after wetting and rinsing. Butylated PVP, Isobutylene/ Ethylmaleimide/ Hydroxyethylmaleimide Copolymer, and Polyquaternium-55 are mentioned as suitable film forming polymers for high internal phase emulsions.

Starches are re-emerging as ingredients in cosmetic formulations because they offer a number of benefits including 'natural' claims, renewability, environmental friendliness, and improved sensory attributes.[27] Starches are used in two forms: soluble starches as colloidal dispersions or granular form (particulate starches). Starches are found in corn, wheat, and rice grains and in potato and tapioca roots. The different kinds of starches offer different properties due to variations in particle shape and size.

The soluble starch form is achieved by heating the granules until the hydrogen bonds between the amylose and amylopectin sections are broken, which releases the starch polymers. This form is beneficial in building viscosity due to hydrodynamic volume exclusion, rheology modification and emulsion stabilization. When films are dried, the soluble starch becomes humidity resistant. It also provides a smooth and conditioned feel to hair and skin.

The granule starch form is achieved by modification of the starch particles so that the hydrogen bonds cannot be broken and the granules cannot rupture. This form delivers a soft and silky feel, aids in oil and/or moisture absorption, and reduces greasiness and/or caking on the skin. The benefits that are achieved are dependent on the source of the starch as well as on the modifications. Rice starch particles are small and spherical, with a tendency to form clumps which causes caking and stickiness. Corn and tapioca starches are larger with good flow properties. The tapioca particles are more truncated which gives a more cushiony and less silky feel.

Starch particles are generally modified by hydrophobic group incorporation along the backbone or by cross-linking. Low friction is important in creating a smooth velvety feel. Long chain hydrophobic modification coupled with ionic cross-linking has been shown to be most effective at reducing friction. Water repellency is important for stability in aqueous formulations. It can also reduce caking and enhance flow properties. Studies have indicated that the presence of a hydrophobe, in general, and more specifically its type are most important for water repellency. Long hydrophobic groups confer water repellency.

Oil absorbance is important for formulations which contain a high amount of oily actives and is also important for providing a soft afterfeel on the skin. Starch particles that were modified using longer hydrophobic groups and ionic cross-linking were shown to provide the best oil absorption.

Heat stability is an important property for the formulating process because many cometic emulsions are made near 80°C. If the starch granules break down before 80°C, they need to be added during the cooling stage in order to avoid property changes. Non-cross-linked starch particles are broken down at very low temperatures. It was found that cross-linking was the main factor in heat stability. Covalently cross-linked particles were most heat stable because the cross-linker was able to prevent the granule from swelling. Ionic cross-links are less strong, however they did increase heat stability over non-cross-linked particles.

Oil-in-water nanoemulsions are liquid/liquid dispersions in which the droplet size is less than 100 nm. These small droplet size emulsions are transparent and the droplets can be stabilized by a shell of emulsifiers, structured as lamellar phase, that completely wraps each droplet. Nanoemulsions are distinctly different from microemulsions insofar as the latter are spontaneously-forming thermodynamically stable micellar phases in which the oil is solubilized. This requires large concentrations of surfactants which make these compositions feel sticky upon application to the skin. Moreover, microemulsions form only in a very limited range of compositions. Nanoemulsions, on the other hand, are similar to regular emulsions by virtue of the fact that they are thermodynamically unstable and the small droplet size is achieved by subjecting the system to enormous shear energies using homogenizers or dispersers. Nanoemulsions often need to be thickened to be acceptable to consumer expectations. There are two conventional ways to thicken a nanoemulsion:

- increase the concentration of the dispersed oil phase. This is usually not an option for the cosmetic formulator because it usually leads to compositions with an undesirable oily or greasy feel
- add conventional polyionic thickeners. But these thickeners usually cause flocculation and/or coalescence of the nanoemulsion and destroy the transparency for which the nanoemulsion was made in the first place.

L'Alloret have recently revealed that nanoemulsions can be thickened without loss of clarity, if the thickener is nonionic[28,29]. Exemplified thickeners are poly (ethylene oxide), hydroxypropyl guar, and hydroxyethylcellulose.

Thickening alpha-hydroxyacid and beta hydroxyacid compositions is another challenge that is not a trivial task for the formulator, especially for low-pH compositions. Complex thickening systems comprising xanthan gum, magnesium aluminum silicate, and polyacrylamide & C13-14 isoparaffin & laureth-7 provide efficient thickening for these systems and they also stabilize cosmetic oil-in-water emulsions.[30] Compositions thickened with the latter crosslinked polyacrylamide have been recognized for improved skin feel, rub-in and absorption characteristics.[31] Copolymers of 2-acrylamido-2-methylpropanesulphonic acid and acrylic acid can thicken strongly acid solutions.[32] If the molecular weight is higher than 6,000,000 and the 2-acrylamido-2-methylpropanesulphonate content is greater than 20 percent, then the polymer may confer yield stress and shear thinning characteristics on the composition. Lorant[33] explained that ultrafine oil-in-water emulsions, prepared by the phase–inversion method,[34] are difficult to 'gel' and stabilize with conventional thickeners but they could be stabilized by a cross-linked poly(2-acrylamido-2-methylpropanesulphonic acid), and this could be achieved at acid pH conditions under which conventional anionic thickeners would exhibit loss of viscosity.[35] Unlike other polyelectrolyte thickeners, these thickeners also resist dramatic viscosity loss in the presence of the UV-A sunscreen1,4-benzene[di(3-methylidene-10-camphorsulfonic)]acid. Acryloyl Taurate/Vinyl Pyrrolidone Copolymer is an efficient thickener and stabilizer for emulsion compositions containing alpha- and beta-hydroxyacids.[36]

Comb copolymers have now been made by polymerizing N-acryloyl taurate with commercially available macromonomers. These polymers provide good thickening and emulsifying properties and clarity[37, 38]

A new class of associative thickeners-thiocarbonate compounds – has been disclosed.[39] These thickeners are multifunctional: they can be used as rheology modifiers, suspending agents, spreading agents, lubricants and film formers. Cationic versions can be used for acidic formulations such as alpha hydroxyacid treatments.

Amphiphilic Block Copolymers as Rheology Modifiers

A block copolymer consists of at least two covalently linked polymer blocks in which the blocks are characteristically comprised of chemically different monomer units. In amphiphilic block copolymers one of the blocks is hydrophilic and the other is hydrophobic. In recent years there has been a flurry of activity in the syntheses and applications of block copolymers[40] following the introduction of techniques for living free-radical polymerization[41,42,43,44]. Prior to the introduction of this technique, block copolymers could be obtained only by anionic or cationic 'living" polymerization or by condensation methods that were applicable only to a small class of monomers. Free radical living polymerization has greatly increased the diversity of monomers that can be assembled into block copolymers.

Amphiphilic block copolymers have the capability to form a sequence of hierarchical morphologies that can be tailored to confer desired mechanical properties on the solid material and specific rheologies on solutions.[45] These polymers readily form micellar aggregates, just like surfactants, except that the dimensions are larger by one to three orders of magnitude.[46, 47] Thus the block copolymer of caprolactone and PEG 550 spontaneously form microemulsions and micelles of size 15 to 125nm.[48] Block copolymer micelles may solubilize oils such as sunscreens.[49] Solutions containing spherical micelles range from being non-viscous to having the properties of associative thickeners[50]; and amphiphilic block copolymers can be nanofabricated[51] to confer a wide range of rheologies. The amphiphilic block copolymer of acrylic acid, methyl acrylate and polystyrene (PAA_{90}-b-PMA_{80}-b-PS_{98}) has been shown to self-assemble in aqueous solution into stable nanostructures having spherical and cylindrical topologies[52] and A-B and A-B-A block copolymers of methacrylic and (A) and styrene (B) form stable micelles that are capable of solubilizing hydrophobic molecules.[53] Block copolymers comprising polycationic blocks and nonionic blocks have been claimed for surface deposition [54,55] and for improved foam retention.[56] The deposits modify the chemical properties of the surface for protection or compatibility; modify its hydrophobic or hydrophilic surface properties; or modify feel or mechanical properties of the substrate from 'two-in-one' products. The examples in this patent application are limited to block copolymers of polyTMAEAMS (methylsulfate [2-(acryloyloxy)ethyl]-trimethylammonium) g/mole) and polyacrylamide.

Easy dispersing triblock B-A-B amphiphilic block copolymers have been disclosed that gel the aqueous phase to obtain a broad range of textures and emulsifying properties and enable the preparation of cosmetic compositions with stable viscosity over wide ranges of pH and temperature.[57] It is also revealed that they make it possible to prepare homogeneous, non-flowing, non-runny products that are soft and slippery when applied and stable on storage.

Living free radical polymerization can also be used to synthesize hyperbranched polymers and there have been claimed to be useful in lipstick formulation.[58]

The thickening of oils, while simultaneously achieving cosmetic efficacy, continues to be a challenge for the formulator. Useful compositions have been thickened by particulates such as clays[59,60] or fumed silica[61,62]; structuring waxes[63], triglyceride gellants such as glyceryl tribehenate[64], and silicone elastomers[65]. Block copolymers such as Hydrogenated Styrene/ Isoprene Block Copolymers are useful as gellants.[66] Guenin *et. al.*[67] reveal that silicones

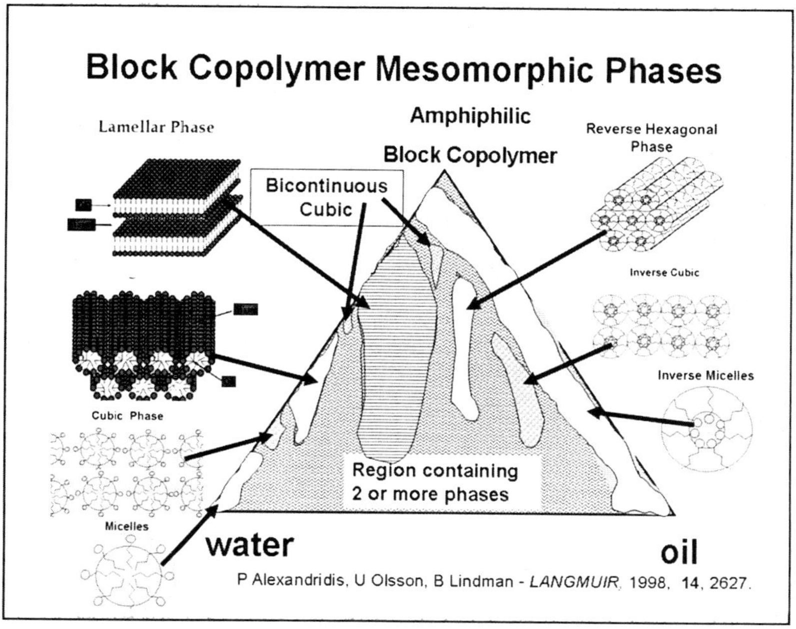

Figure 1: A representative phase diagram of an amphiphilic polymer showing the wide range of morphologies that can occur in oil/water systems with these polymers. (Reproduced from HAPPI *April 2006. Copyright 2006 R. Y. Lockhead.)*

thickened by elastomers tend to have unacceptably high viscosities and they tend to leave a film on the skin that occludes the sweat glands. In order to overcome these deficiencies, they propose a thickener comprising dimethicone cross-linked with an alpha-omega diene in combination with polyethylene beads.

Facial masks represent a type of solid 'emulsion'. There are two types of masks, namely treatment masks and removal masks. Treatment masks adhere very weakly to the skin and are easily removed. These are used for delivering benefit agents such as moisturizing agents to the skin through a wet (usually aqueous) film. Removal masks are designed to firmly adhere to the skin to remove dirt, sebum clogs and excess cornified cells from the surface and pores of the skin.[68] Treatment masks are made from poly (vinyl alcohol) of gels and the inclusion of treatment ingredients such as emollients and oils can alter feel of the mask on the skin. The inclusion of thickeners enables the formulator to include treatment ingredients without sacrificing skin feel attributes.[69] A suitable thickener for this purpose is Sodium Polyacrylate (and) Hydrogenated Polydecene (and) Trideceth-6 .

Hybrid Polymers that Thicken and Form Films

Polymers have recently been developed to simultaneously confer thickening and film-forming capabilities. These polymers having the INCI name of Polyacrylate-14 contain an acidic vinyl monomer (for example methacrylic acid), a nonionic monomer (for example ethyl acrylate), an associative monomer (for example a long chain alkyl ethoxylated methacrylate) , and optionally a semi-hydrophobic monomer (for example an alkyl polyethoxylated methacrylate) and a cross linker (for example trimethylolpropanetriacrylate). These new polymers also provide the added advantage of being fixatives that can be restyled by merely spritzing with water and regrooming. Easy restyling products have become an important segment in a market that is moving towards water-soluble, low-VOC fixatives. For example, Dehydroxanthan gum provides both thickening and hair fixative properties with high holding even in humid conditions and the added advantage of restylability by merely spritzing with water and combing or brushing.

Polymers in Hair Coloring

Polymers play several roles in the coloring of hair. They act as thickeners for oxidation dyes and conditioners and hair-protectors to mitigate the harsh effects of coloring. The thickeners include a range of associative thickeners (anionic, cationic, amphoteric, nonionic, and hydrophobically modified alkali swellable) as well as hydrophobically modified ethoxylated urethanes. Alkylmethicones and aminosilicones play a role in color retention. Non-associative thickeners help in achieving hair lightening.

Thickeners for Hair Dyes

Oxidation dyes are made up of dye precursors (such as *ortho-* and *para-*phenylenediamine, or *ortho-* and *para*-amino phenols). The dye precursors are colorless compounds that can penetrate the hair fiber and in the presence of oxidizing agent, they condense to become colored compounds. The color can be modified by including "couplers" such as meta-phenylenediamine.

The challenges that must be overcome with two-part dye systems are: prevention of separation during storage; enhancing the slow diffusion of dye precursor onto the hair fibers; and ensuring efficient rinseability of the colorant after application.[70] Casperson listed the following conditions of thickened oxidation dye compositions:[71]

1. The formulations must be stable to insure a reasonable shelf life.
2. The compositions formed by mixing the lotion and developer must have rheological properties to allow the applied composition to readily

distribute the dye throughout the hair mass and avoid dripping or running from the hair during the color development period.

3. The dye mixture, as applied to the hair, should allow rapid diffusion of the dye precursors from the dye mixture into the hair fiber.
4. The thickened mixture should be readily rinseable from the hair with water.
5. The mixture should contain conditioning agents that leave the hair easy to detangle while wet and should feel smooth and be readily managed when dry.
6. The lotion and developer should preferably, but not necessarily, have comparable viscosities in order to facilitate mixing.
7. The dyeing effect should be rapid, with a dyeing time preferably under 30 minutes.

The oxidation dye compositions are usually thickened to localize the dye action on the hair and to prevent the dye composition from flowing down the face and into the eyes. Originally, thickeners such as Carbomers,[72] hydroxyethylcellulose, or low HLB surfactants were used. It has been reported that oxidation dye compositions thickened with these traditional thickeners result in a dull shade on the hair and as a result, high concentrations of dyes are required to obtain intense colors.[73]

Oxidation dyeing is a complex process the structure of the thickener can alter the rate and mechanism of diffusion of dye precursor and condensate into the hair. Therefore, it is interesting to consider the structure of the thickeners with their reported performance in oxidation dyeing. Associative thickeners are increasingly being used for this purpose. There are several classes of associative thickeners that should be considered:[74]

- ***Anionic associative polymers:*** Steareth-10 allyl ether/acrylates copolymer is the INCI nomenclature for terpolymers of methacrylic acid ethyl acrylate and steareth-10 allyl ether. (Acrylates/Steareth-20 methacrylate copolymer is the INCI nomenclature for methacrylic acid/ethyl acrylate/stearyl methacrylate oxyalkylenated terpolymer. Acrylates/C10-C30 alkyl acrylate crosspolymer are cross-linked copolymers of acrylic acid and C10-C30 alkyl (meth)acrylates. Maleic anhydride/C30-C38 alpha olefin/alkyl maleate terpolymers and Polyacrylate-3 (a copolymer of methacrylic acid, methyl methacrylate, methylstyreneisopropyl isocyanate and PEG-40 behenate monomers) are further examples of anionic associative thickeners.

- ***Cationic associative polymers:*** These include associative polyurethanes, such as polyureas or polythioureas, prepared by reaction of selected amines, diisocyanates and hydrophobic compounds that are capable of coupling to form telechelic end groups on the polymer.

Polyquaternium-24 is the quaternized ammonium salt of hydroxyethylcellulose reacted with a lauryl dimethyl ammonium substituted epoxies. PG hydroxyethylcellulose coco-dimonium chloride, PG hydroxyethylcellulose lauryldimonium chloride, and PG hydroxyethylcellulose stearyldimonium chloride are quaternary ammonium salts of hydroxyethylcellulose.

- ***Amphoteric associative polymer:*** One example is acrylic acid/(meth)acrylamidopropyl-trimethylammonium chloride/stearyl methacrylate terpolymers.

- ***Nonionic associative polymers:*** Cetyl hydroxyethylcellulose is the cetyl ether of hydroxyethylcellulose, Hydroxyethyl ethylcellulose is the ethylene glycol ether of ethylcellulose. Alkyl hydroxypropyl guars are the alkyl ethers of hydroxy propyl guar. VP/Hexadecene copolymer and VP/Eicosene copolymer are copolymers of N-vinyl pyrrolidine with hexadecane and eicosene, respectively. PEG-150/stearyl alcohol/SMDI copolymer is a copolymer of PEG-150 saturated methylene diphenyldiisocyanate and stearyl alcohol monomers. PPG-l4 palmeth-60 hexyl dicarbamate is the carbamic acid diester of the polypropylene, polyoxyethylene ether of the fatty alcohols derived form *Elaeis guineensis* (palm) kernel oil. Polyether-1 is a copolymer of PEG-180, Dodoxynol-5, PEG-25 tristyrylphenol, and tetramethoxy-methylglycouril monomers. The aminoplast skeleton of these thickeners offers enhanced stability.

- ***Hydrophobically modified alkali swellable associative polymers:*** Pohl *et al* have reported that acrylates/steareth-20 methacrylate copolymer was capable of thickening hydrogen peroxide solution and this made it possible to prepare a 2-part system in which the acrylates/steareth-20 methacrylate copolymer, hydrogen peroxide and dye precursors in one part were thickened by adding a second part consisting of a base such as ammonium hydroxide.[75] However, anionic acrylate thickeners tend to form complexes with the quaternary ammonium surfactant conditioners in the composition. Also, formulations thickened with simple aqueous polyacrylate are difficult to rinse from the hair and it is difficult to formulate to a consistent viscosity with Acrysol ICS-1. Casperson et al reported that the use of the acrylates/steareth-20 methacrylate copolymers overcame these problems. Acrylates/steareth-20 methacrylate copolymer is a hydrophobically-modified alkali-swellable (HASE) thickener. In this case there are two mechanistic components that work jointly to produce the observed thickening. (1) The first component is polyion swelling, which occurs when the polymer is neutralized with base and becomes a polyelectrolyte; the resulting polyion swells as a result of mutual ionic repulsion of the carboxylate ions that are covalently attached to the polymer chain. (2) In the second mechanistic component, swollen thickener molecules are hydrophobically associated to form an overall network structure. The hydrophobic associations between the macromolecules are easily disrupted by shear and are reformed in a new conformation when the shear force is ceased. The hydrophobic associations confer useful properties. For example, the efficiency of a molecular "network" is obtained without the elastic recoil that would normally be connected with crosslinked elastomers. However, steric hindrance of the polymer chains results in a significant proportion of the hydrophobic groups being isolated and unassociated in the aqueous environment. This leads to the interesting observation that addition of surfactant causes first an increase in measured viscosity followed by a viscosity decrease when the surfactant concentration exceeds the critical micelle concentration. Such behavior is observed because co-micellization between the "isolated" polymer hydrophobes causes an increase in the number of "crosslink" sites.[76] However, when sufficient micelles are present, the stoichiometry of co-micellization favors at most only one polymer hydrophobe per micelle and mutual repulsion between the micelles causes disruption of the network. Hydrophobically-modified copolymers of acrylamidopropane sulphonic acid or salt and methacrylate esters of ethoxylated or propoxylated fatty alcohols when incorporated with hydrogen peroxide in a developer formula and used with an oxidation dye precursor compound have been claimed to achieve improved stability against separation that leads to an enhancement of hair coloration. The polymer in this case is ammonium acryloyldimethyltaurates/beheneth-25 methacrylate crosspolymer.

- ***Hydrophobically modified ethoxylated urethanes (HEUR)*** enhance color intensity from thickened oxidation dyes. The HEUR thickeners are ABA triblock copolymers in which the A blocks are hydrophobic and the B blocks are hydrophilic poly (alkoxylate). HEURs are also available as graft/block copolymers with comb architecture. These polymers aggregate into makes and, in the micelle, these molecules are bent into "hair-pin" confirmation with the hydrophobic ends in the micelle core and the hydrophilic "middle" interacting with water in the micelle corona. A network is formed by some HEUR molecules that stretch between and link the micelles (Figure 3).[77] Comb-type HEURS show low aqueous solution viscosity, but they do offer better network formation and co-micelle association.[78] Systems thickened by HEURS display pronounced shear thinning when the shear stress exceeds a critical value that corresponds to the stress that is necessary to extract the linking molecules from the micelles.[79] This model is supported by the fact that these systems can display a single relaxation time that corresponds to the process of the hydrophobic group being extracted from the micelle.[80] The addition of even low concentrations of conventional surfactants to HEUR-thickened systems results in a dramatic loss of viscosity due to competition for residence of the micelle core between the surfactants and the HEUR linking molecules.

- Aminoplast thickeners function by similar rheological mechanisms as HEUR thickeners. Aminoplast reagents are the aldehyde reaction products of melamines, ureas, benzoguanamines, and glycolurils.[81] The aminoplast thickeners are taught to cause bleaching[82] and dyeing[83] formulations to remain localized at the position of application on the hair so that they do not run over the face. They are claimed to leave the hair less coarse than the currently used thickeners. The polymers described for this application include PEG-180/octoxynol-40/TMMG copolymer, PEG-180/laureth-50/TMMG copolymer, and polyether-1 .

Once the hair has been dyed with either a direct dye or an oxidative dye, the color must be maintained. Color retention can be enhanced by the addition of silicones during the hair coloring process. Longer-lasting hair color retention and volumizing has been claimed by the addition of alkylmethicones either before, during or after the hair has been dyed.[84] Suitable alkylmethicones are hexylmethicone, octylmethicone, caprilylmethicone and stearoxytrimethyl-silane. Color retention is usually evaluated by half head tests in which the color was measured at root, middle and tip of hair by a chromometer.[85]

Pretreatment of hair with an aminosilicone is claimed to confer better color retention from direct dyes and oxidative dyes.[86] These compositions confer water contact-angles greater than 90° on the hair surface measured by the Wilhelmy plate method using the hair fiber as the plate. Adsorption of silicones to hair has been claimed to be too weak to provide lasting protection. On the other hand, covalent binding of silicones to hair causes damage to and weakening of the fiber. However, the covalent attachment of trimethylpyridinium "molecular hooks" to the silicone chain is claimed to produce a silicone that adsorbs to the hair and is durable, while maintaining intact hair structure.[87]

Hair Lightening: Hair can be lightened with bleaching agents such as hydroquinone but hydroquinone is known to be cytotoxic to melanocytes. It has been revealed that hair lightening can be achieved by application of compositions containing a fluorescent dye and a cationic polymer and it appears that these compositions give improved fixing of the dye on the hair surface. The cationic polymer must have a charge density of at least 1 meq/g.[88] The fluorescent dye absorbs light in spectral wavelengths between 360 and 760 nanometers and converts it into light of longer wavelength that is emitted in the visible region. This is distinguished from optical brighteners, which absorb only ultraviolet light (200 to 400 nanometers).

It has also been claimed that hair lightening can be achieved by application of compositions containing a fluorescent dye and a non-associative thickener. Suitable non-associative thickeners can be found among crosslinked acrylic acid homopolymers, nonionic guar gums, and dimethylaminoethyl methacrylate homopolymers and copolymers quaternized with methyl chloride.

Conditioning Polymers

Hair Conditioning

Conventional conditioner formulations are based upon lamellar gels or emulsions using either ceto-stearyl trimethylammonium chloride or distearyldimethylammonium chloride as cationic surfactants and ceto-stearyl alcohol as co-surfactant. These products form a gel matrix that confers conditioning benefits from rinse-off products. They have been the basis of hair conditioners for the last half-century and they do provide excellent detangling, wet- and dry-combing, and good anti-static properties, but they can leave the hair feeing lank and greasy, and they give a long-lasting slippery feel during rinsing and this is perceived by some consumers as unclean hair feel.

Cationic polymeric conditioners can improve wet combability and ameliorate electrostatic charging of the hair (manifested by flyaway). For example, Polyquaternium-10 (which is cationic hydroxyethylcellulose), Polyquaternium-6 (polydiallyldimethylammonium chloride homopolymer), Polyquaternium-7 (which is a copolymer of diallyldimethylammonium chloride and acrylamide), are found in conditioner formulations.

There is a need for clear conditioning compositions with suitable "conditioning" rheology conferred from something other than a gel matrix (defined as having a Brookfield RT viscosity of 5,000 to 20,000 cps at 2 sec^{-1}) and rinse-off to leave a clean feel while depositing sufficient material to confer benefits such as softness and reduced tangling of wet hair, and good wet-combing. Crystal clarity in the formulation is also desired as a visual cue to the consumer. Such conditioners should not weigh down the hair and should offer the opportunity to volumize fine hair. A recent patent application[89] claims that these benefits can be achieved from aqueous-based conditioners comprising a cationic crosslinked thickening polymer and nonionic, cationic surfactants or mixtures of these. It is notable that these claimed conditioners contain no water-insoluble high-melting-point oils or waxes. The cationic crosslinked polymers are copolymers of dimethylaminomethacrylate and acrylamide crosslinked with methylene-*bis*-acrylamide (Polyquaternium-32). A homopolymer of quaternized dimethylaminoethylmethacrylate crosslinked with methylene-*bis*-acrylamide is especially preferred (Polyquaternium-37). Similarly, a patent has been awarded

Table 3: Examples of Conditioning Polymers

Chitosan	
Cocodimonium Hydroxypropyl Hydrolyzed Collagen	
Cocodimonium Hydroxypropyl Hydrolyzed Hair Keratin	
Cocodimonium Hydroxypropyl Hydrolyzed Keratin	
Cocodimonium Hydroxypropyl Hydrolyzed Wheat Protein	
Cocodimonium Hydroxypropyl Oxyethyl Cellulose	
Steardimonium Hydroxyethyl Cellulose	
Stearyldimonium Hydroxypropyl Hydrolyzed Oxyethyl Cellulose	
Guar Hydroxypropyltrimonium Chloride	
Starch Hydroxypropyltrimonium Chloride	
Lauryldimonium Hydroxypropyl Hydrolyzed Collagen	
Lauryldimonium Hydroxypropyl Hydrolyzed Wheat protein	
Stearyldimonium Hydroxypropyl Hydrolyzed Wheat Protein	
Polyquaternium-4,	
Polyquaternium-10,	Cationic hydroxyethylcellulose
Polyquaternium-24	Hydrophobically modified cationic hydroxyethylcellulose
Poly(methacryloxyethyltrimethylammonium methosulfate)	
Poly(N-methylvinylpyridinium chloride)	
Onamer M (polyquaternium-1), PEI-1500 (Poly(ethylenimine)	
Polyquaternium-2	
Polyquaternium-5-poly(acrylamide-*co*-methacryloxyethyltrimethylammonium ethosulfate)]	
Polyquaternium-6 poly(dimethyldiallylammonium chloride)	
Polyquaternium-7	poly(acrylamide-*co*-dimethyldiallylammonium chloride)
Polyquaternium-8	
Polyquaternium-11	[poly-(N-vinyl-2-pyrrolidone-methacryloxyethyltrimethylammonium ethosulfate)]
Polyquaternium-16 [Co(vinyl pyrrolidone-vinyl methylimidazolinium chloride)	
Polyquaternium-17	
Polyquaternium-18	
Polyquaternium-22	poly(sodium acrylate – dimethyldiallyl ammonium chloride)
Polyquaternium–27	
Polyquaternium-28_	polyvinylpyrolidone-methacrylamidopropyltrimethylammonium chloride)
Polyquaternium-31	Poly(N,N-dimethylaminopropylacrylate-N-acrylamidine-acrylamide-acrylamidine-acrylic acid-acrylonitrile) ethosulfate
Polyquaternium-39	poly(dimethyldiallylammonium chloride – sodium acrylate – acrylamide)

Table 3: *Continued.*

Polyquaternium-43
poly(acrylamide-acrylamidopropyltrimoniumchloride-2-acrylamidopropyl sulfonate-DMAPA)
Polyquaternium-44
Poly (vinyl pyrrolidone-*co*-imidazolinium methosulfate)
Polyquaternium-46
Poly (vinylcaprolactam-vinylpyrrolidone-imidazolinium methosulfate)
Polyquaternium-47
poly (acrylic acid-methacrylamidopropyltrimethyl ammonium chloride – methyl acrylate)
Polyquaternium-53
Polyquaternium-55
poly(vinylpyrrolidone-dimethylaminopropylmethacrylamide-lauryldimethylpropylmethacrylamido ammonium chloride)
PVP/Dimethylaminoethyl Methacrylate Copolymer
VP/DMAPA Acrylate Copolymer
PVP/Dimethylaminoethylmethacrylate Polycarbamyl Polyglycol Ester
PVP/Dimethiconylacrylate/Polycarbamyl Polyglycol Ester
Quaternium-80 (Diquaternary polydimethylsiloxane)
Poly(vinylpyrrolidone-*co*-dimethylamidopropylmethacrylamide)
VP/Vinyl Caprolactam/DMAPA Acrylates Copolymer
Amodimethicone
PEG-7 Amodimethicone
Trimethylsiloxyamodimethicone
Ionenes
Poly(adipic acid-dimethylaminohydroxypropyldiethylenetriamine)
poly (adipic acid-epoxypropyldiethylenetriamine) (Delsette 101)
Silicone Quaternium-8
Silicone Quaternium-12

for a process of conditioning hair by the application of a crosslinked cationic thickening polymer.[90] The polymer comprises N-vinylimidazoline or a derivative, an N-vinyllactam such as N-vinylpyrollidone, and a short chain alkyl (meth)acrylate.

The application of cross-linked cationic polymers as conditioners could constitute a new approach to the challenge of hair conditioning.

Polyurethane conditioners are claimed to confer enhanced conditioning.[91] These polymers are synthesized by reacting trialkanolamine (preferably triethanolamine) with a hydroxy fatty acid (ricinoleic acid, lactic acid or salicylic acid) to make a mono-, di- or triester, then reacting with a di-isocyanate to produce a polyurethane with tertiary amine functionality, then quaternizing the tertiary amines. The benefits claimed give better adhesion to hair and skin than conventional cationic conditioners and better sheen, wet combing, dry combing, antistatic and conditioning properties.

Leave-on silicone conditioners specifically targeted to non-shampoo applications have been disclosed.[92] These contain emulsified vinyl-terminated silicones applied in combination with a conventional cationic conditioner. A preferred product type is a mousse. The significance of this claimed invention rests on the basis that these silicone block copolymers can achieve excellent conditioning at relatively high viscosities (100 $KPa.s^{-1}$). Aqueous foaming shampoo compositions containing emulsified vinyl silicone polymers are described in another application.[93]

Conditioning Polymers in Hair Straightening Applications: The two main processes for relaxing or straightening hair are hair treatment with a reducing agent to cleave of the disulphide cystine bridges (S--S) within the hair structure, and treatment of stretched hair with a strong alkaline agent.

Repeated relaxation treatments can cause significant hair damage, to both the cuticles and the cortex. The damage can be assessed by measuring the porosity of the hair and the porosity of the keratin fibers can be measured by fixing 2-nitro-para-phenylenediamine at 0.25% in an ethanol/buffer mixture (10/90 volume ratio) at pH 10 at 37°C for 2 minutes.[94] Cationic and amphoteric polymers such as Polyquaternium-6, Polyquaternium-7 and Polyquaternium-39, added to hair relaxer formulations, mitigate this degradation of the hair structure. Also, the inclusion of high molecular-weight (>106 g/mole) copolymers of acrylamide and diallyldimethylammonium chloride, acryloyloxytrimethylammoniumchloride, or acryloyloxyethyldimethylbenzylammonium chloride in the relaxing formula results in significant reduction in the hair structural damage caused by alkaline relaxation.[94] The polymers are synthesized by precipitation polymerization in salt solution and delivered to the substrate as a particulate dispersion in a concentrated salt solution.

Cleansing

Soap: Soap has been used as a cleanser for thousands of years and its use became common during the industrial revolution. However, solid bar soap has a relatively slow time for dissolution and solubilization. This can be improved by providing the product as a liquid soap but the high water and solvent content of this product form results in increased costs of manufacturing and shipping for this commodity product. Recently, a solid, rapidly dissolving soap composition, suitable for pre-dose dispensing, has been disclosed.[95]. The sheet-like product is prepared by combining soap with a water-soluble film forming polymer and introducing air bubbles. Suitable polymers are listed as polyamides, polyacrylates, polyamino acids, polyvinyl acetate, polyvinyl alcohol, polyethylene glycols, polyvinylpyrrolidones, pullulan, alginic acid, starch, cellulose and cellulose derivatives. Only poly(vinylalcohol) is exemplified in the patent application. Consistent with storage and dispensing, the product is designed to rebound to its original shape, like a sponge, after deformation under a slight external force but to show permanent deformation under greater forces.

The formulation of personal cleansing compositions often involves a compromise because the best cleansing and foaming detergents (such as linear alkyl benzene sulfonates or alkyl sulfates) are irritating to skin and mild surfactants are usually poor cleansers and low foamers. Detergents containing polymeric suds stabilizers and diamines are touted for improved mildness and skin feel because they enhance cleansing and foaming and offer low-level surfactant compositions with acceptable performance.[96] The suds stabilizers are cationic polyelectrolytes, proteins or zwitterionic polymers. The cationic polyelectrolyte for this purpose have a charge density in the range 0.05 to 3 cationic units per 100 daltons of polymer molecular weight.[97] For example, poly(dimethylaminoethylmethacrylate–*co*–dimethylacrylamide), poly(dimethylaminoethylmethacrylate–*co*–acrylic acid), poly(dimethylaminoethylmethacrylate–*co*–methacrylic acid), poly(dimethylamino-ethylmethacrylate–*co*–methacrylic acid–*co*–acrylic acid), poly(dimethylamino-ethylmethacrylate–*co*–methacrylic acid–*co*–dimethylacrylamide, polypeptides containing lysine, alanine, glucosamine and tyrosine of molecular weight 52,000 daltons, and lysozyme.

It is interesting to consider the possible mechanism of foam stabilization of anionic surfactants by cationic polymers. The stability of foam films depends upon the forces that act across the two faces of the film. Van der Waals forces attract the film surfaces toward each other and repulsive forces (such as ionic repulsion) hold the surfaces apart and stabilize the foam. The repulsion causes a hydrostatic pressure inside the film and this pressure is often called the 'disjoining' pressure because it is the pressure that holds two adjacent gas bubbles apart in the foam. High repulsive disjoining pressures correspond to stable foams.[98] When polyelectrolytes and surfactants of opposite charge are present in foam, the foam may exhibit enhanced stability. There are two distinct methods of stabilization depending upon the surfactant concentration. Specifically the mechanisms depend upon whether the surfactant concentration is below or above the critical aggregation concentration (CAC).[99] Below the CAC the film is initially stabilized by electrical double layer repulsion between the surfactant head group on opposite film faces, and the polyelectrolyte 'floats' free in the aqueous fluid within the film. (figure 2). Surfactant films drain due to gravitational and capillary forces. In particular, the curvature of the film at the Plateau Border of the foam causes a difference in Laplace Pressure between the film lamellae and the Plateau Border ($\Delta P = 2\gamma/r$, where ΔP is pressure difference, γ is the surface tension and r is the radius of curvature) and this causes fluid to flow from the lamellae into the Plateau Border. Drainage of the film causes the lamellae to thin and eventually a point is reached when the *Van der Waals* attraction between the film faces exceeds the disjoining pressure. At this point a simple surfactant-water film would switch instantaneously and 'squeeze out' the remaining fluid in the lamellae to become either a stable black film or unstable lamellae that would quickly break. (A black film is one which results from complete removal of the mobile aqueous phase between the surfactant faces). However, in the presence of an oppositely-charged polyelectrolyte, the polymer can 'bridge' between the film faces and hold the lamellae apart resulting in enhanced foam stability in the drained foam. (Figure 2) Above the CAC, surface-active polymer-surfactant complexes (coacervates) are formed and these adsorb at the interface lowering the surface tension, enabling foam formation and stabilizing the foam. The coacervates phase-separate from solution and they are adsorbed in the soap film as discrete heterogeneous domains that appear as thick 'spots' in the film.[99] When the film drains, the polymer-surfactant complexes can form a gel-network that prevents film collapse and rupture. Therefore, a more stable foam results.

Cationic polyelectrolytes can cause anionic surfactants to self assemble into liquid crystals that phase-separate from solution.[100] It has been revealed that the cationic polyelectrolyte in these applications must have a non-crosslinked flexible backbone to permit the polyions to form the closely associated layers of polymer and self-assembled surfactant that comprise the phase-separated liquid crystalline phases.

When cleansing products based upon such compositions are applied to the skin, these liquid crystals can deposit leading to skin conditioning. Moreover, it was discovered that such liquid-crystal compositions offered an improved capability to suspend particles and paradoxically, better deposition of the particles to the substrate (skin or hair) during rinsing. Particle deposition is desired in applications such as pigments or coloring agents, opacifiers, pearlescent agents, feel modifiers, oil absorbers, skin protectants, matting agents, friction enhancers, slip agents, conditioning agents, exfoliants, odor absorbers, cleaning enhancers, and particulate active ingredients including antiperspirant agents, anti-dandruff agents, antimicrobials, antibiotics, and sunscreens.

2-in-1 Cleansing and Moisturizing: The ideal situation to deliver moisturization to the skin is in the shower or bath. However, deposition is challenging in these situations when there is a tendency for the moisturizer to be rinsed away form the skin. There have been several attempts to meet the challenge. For example:

- high internal phase emulsions break easily on the skin and, if they are substantially free of surfactant, satisfactory deposition of moisturizer can be achieved[101]
- delivery of deodorants from spheroidal networks comprising wetting agents and emulsifiers such as stearic acid, cetyl alcohol, glyceryl monostearate and stearyl alcohol[102, 103].
- patterned multi-phase systems that comprise a a cleansing phase, a benefit phase, and a non-lathering structured aqueous phase.[104]

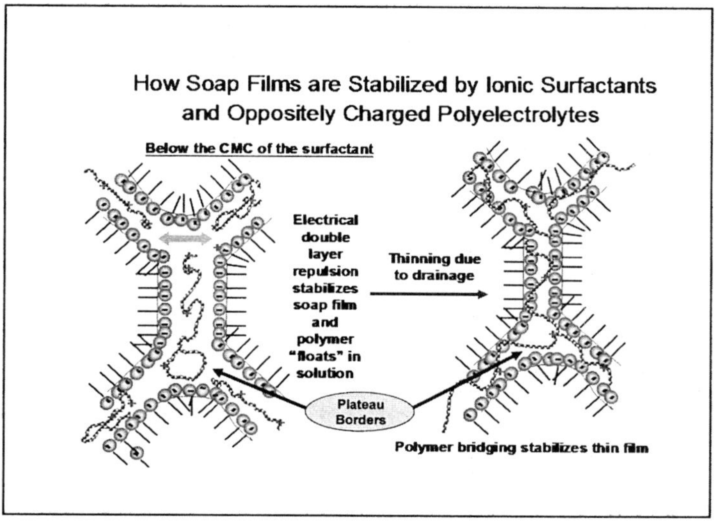

Figure 2: A schematic illustration of foam stabilization by cationic polymer below the surfactant CMC. (Reproduced from HAPPI *April 2006. Copyright 2006 R. Y. Lockhead.)*

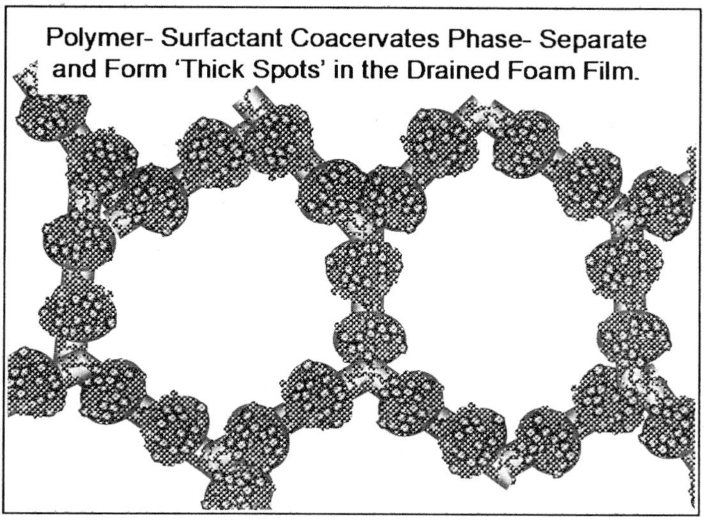

Figure 3: A schematic illustration of the location of polymer-surfactant coacervate in the drained foam. (Reproduced from HAPPI *April 2006. Copyright 2006 R. Y. Lockhead.)*

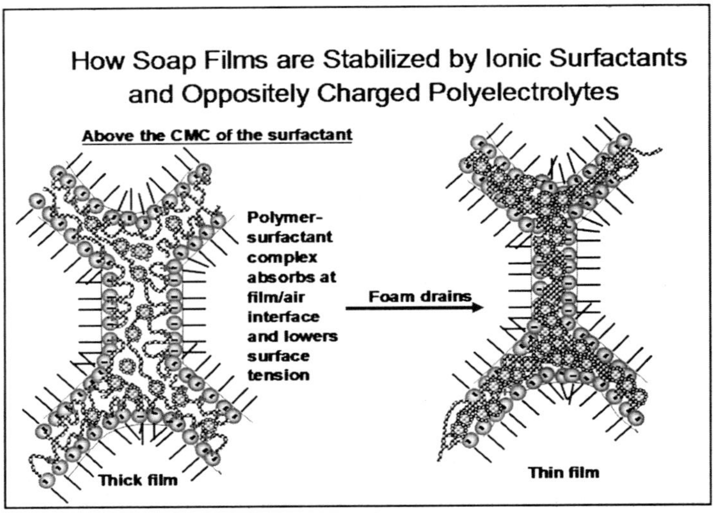

Figure 4: A schematic illustration of foam stabilization by cationic polymer above the surfactant CMC. (Reproduced from HAPPI *April 2006. Copyright 2006 R. Y. Lockhead.)*

- oil-in-water emulsions in which the oil is structured by a stable network of finely divided particles such as silicas, clays or trihydroxystearin.[105] The structured oil has a high low-shear viscosity that favors adhesion and retention on the skin, and the network yields under stress when rubbed in. The viscosity of the system can be built with a polymeric stabilizer[106] such as the naturally-derived hydroxypropyl starch phosphate. This is a pre-gelatinized esterified form of starch that fulfills several roles in skin care products, such as emulsion stabilization, improved sensory quality and viscosity build. Hydroxypropyl Starch Phosphate can enhance formulation aesthetics as a dispersed starch. This starch polymer gives body to formulations and leaves the skin feeling soft and it is a nonionic polymer that is compatible with most skin care ingredients. This polymer is easy to use and it can be added at virtually any point in the production process This starch derivative also addresses the desire of consumers of cosmetics and toiletries across the world to increasingly expressing a preference for products that contain naturally derived ingredients. Such ingredients are perceived favorably with respect to health, safety and environmental responsibility. However, consumers are not willing to sacrifice product performance to satisfy a preference for natural ingredients. Today's consumers actually demand products with increasingly sophisticated performance. This creates a formidable challenge for cosmetic scientists - to develop new products containing naturally derived raw materials that meet or exceed performance expectations. Hydrophobically- modified pre-gelatinized crosslinked starches are polymeric emulsifiers their yield emulsions that are salt-tolerant and do not exhibit tackiness.[107] Corn Starch Modified, provides unique smooth, velvety feel benefits. One of the more distinctive properties of this unique polymer is its ability to impart a pleasantly light, soft feel to a range of product forms (lotions, creams and pressed powders) both during application and after dry-down. It has the ability to mitigate greasiness, even in heavy formulations.

- Moisturizing composition for adolescents usually require a sebum control agent. Liquid crystal /gel networks containing a sebum absorbing agent, preferably Aluminum Starch octenyl succinate (and) Acrylates Copolymer (and) Magnesium Carbonate, have been proposed for this purpose[108]. Adding between 1% and 3% of this hydrophobically –modified starch polymer allows for a matt finish on skin, reducing unwanted shine.

Conditioning Shampoos: The origin of conditioning shampoos can be traced to the Balsam Shampoos of the 1960s followed by the introduction of polyquaternium-10 and the groundbreaking work and scientific insight of Des Goddard[109] in the 1970s and 1980s in which he introduced the concept of polymer-surfactant complex coacervates that phase-separate and deposit on the hair during rinsing. The original successful conditioning shampoos of the 1970's were based upon the formation of coacervates of cationic polymer (such as Polyquaternium-10 or Guarhydroxypropyltrimonium chloride) and anionic surfactants. Guar is a galactomannan and recently a new cationic galactomannan hydrocolloid, cationic cassia, has been introduced.[110]

During the 1980s and into the 1990s, complex coacervates were employed as vehicles for the deposition of silicone on hair as conditioning shampoos evolved.[111] This trend continues today and the two original polyelectrolytes – Polyquaternium-10 and Guar hydroxypropyltrimonium chloride – continue to be the predominant conditioning ingredients in these shampoos. The mechanism of conditioning shampoos depends upon the formation of polymer/surfactant coacervates that phase-separate during rinsing. Polyions in aqueous solution are surrounded by an electrical double-layer of counterions and the location of the counterions with respect to the polyion is determined by a balance between chemical potential and electrochemical potential, called the "Donnan Equilibrium." Surfactant ions contain a large hydrophobic group and this makes them intrinsically less soluble in water than inorganic ions such as chloride or bromide. When surfactant ions interact with an oppositely charged polyion, they bind strongly and displace the water-soluble inorganic ions from the polyion; that is, they ion-exchange. Once the surfactant ions bind, hydrophobic interaction between the hydrophobic surfactant tails causes the polymer-surfactant complex to phase separate. If the surfactant concentration is sufficiently high to form micelles or hemi-micelles along the polyion chain, then the polyion is resolubilized. Conditioning shampoos are formulated within the range of surfactant concentrations that correspond to this solubilized regime. If this solubilized complex (shampoo) is diluted to a concentration in the vicinity of the critical micelle concentration, then the complex coacervate phase-separates. The separated phase is deposited on the hair during rinsing.

Interest in these systems has recently been revived due to advances in our knowledge of the nanoscience of self-assembling systems and polymer-surfactant coacervates. The trend has been driven commercially by the desire to enhance deposition of "actives" and benefit agents from cleansing compositions during the rinsing stage. Conditioning shampoos are expected to confer the wet hair attributes of hair softness and ease of wet-combing, and the dry hair attributes of good cleansing efficacy, long-lasting moisturized feel, and manageability with no greasy feel. The development of free-radical living polymerization offers the opportunity to greatly extend the variety of polymers, especially block copolymers, which will be available to the cosmetic formulator. Block copolymers comprising polycationic blocks and nonionic blocks for surface deposition[112,113] and for improved foam retention[114] have recently been claimed that are desired to deposit on has to modify the chemical properties of the surface for protection or compatibility; to modify hair's hydrophobic or hydrophilic surface properties; or to modify feel or mechanical properties of the substrate from 'two-in-one' products. The examples in this patent application are limited to block copolymers of polyTMAEAMS (methylsulfate [2-(acryloyloxy)ethyl]-trimethylammonium) g/mole) and polyacrylamide.

Recent research efforts have been directed towards optimizing polymer and surfactant structure and formulation variables to seek enhanced benefits; including other components to enhance the conditioning benefits; improving the clarity of conditioning shampoos, and seeking new methods of conferring conditioning from shampoos that do not contain a cationic polymer.

Adding other components to enhance conditioning: Silicones have become standard ingredients in many conditioning shampoos for the smooth, silky hair feel that they confer. Volatile cyclic siloxanes confer the desired silky initial feel but these materials are difficult to formulate in consistent homogenous formulations and, according to O'Brien[115], they tend to spread uncontrollably over hair and skin. Polymeric silicone gels formed in volatile silicones provide both the initial silky feel and a high-viscosity, smooth feel upon drying[116,117,118]. These compounds are synthesized by hydrosilylation of unsaturated hydrocarbons. Recently new classes of silicone gels suitable for hair products have been disclosed; these are produced by hydrosilylation of oxirane ring compounds such as epoxides[115], or by forming branched molecules with a silicone core and hydrocarbon branches, or networks formed from these branched units.[119]

The deposition of silicone conditioning components from a detersive, cleansing system is somewhat paradoxical because the detersive system is designed to remove oil, grease, dirt, and particulate material from the very hair upon which the conditioning agent has to be deposited and as a consequence large amounts of the expensive conditioning silicone can be rinsed away rather than deposited on the hair. In cosmetic formulation circles, it is well known that cationic polymers can enhance the deposition of benefit agents.[120,121]. Enhanced deposition is claimed from compositions that contain two cationic polymers (guar hydroxypropyltrimonium chloride and acrylaminopropyltrimonium chloride/acrylamide copolymer).[122] It has now been disclosed that deposition can be enhanced by pre-emulsification of hydrophobic benefit agents with cationic emulsifiers, prior to addition to cationic polymer/anionic surfactant compositions.[123]

A recent patent application discloses that shampoos containing more than one cationic conditioning polymer and a quaternary silicone give more uniform deposition on hair than standard shampoos based on Polyquaternium-10 as the sole conditioning polymer.[124] According to claims in the invention, the conditioning polymer "cocktail" comprises poly(acrylamide-co-acrylamidopropyltrimonium) chloride, guar hydroxypropyltrimonium chloride, and silicone quaternium-13. The claims are based upon multiple testing and analysis, namely:

- A Multiple Attribute Consumer Assessment Study that measured the attributes of cleaning, wet-comb, dry-comb, hair softness, lather amount and creaminess.
- Secondary Ion Mass spectrometry to detect silicon on the hair surface. This method revealed that a standard commercial shampoo concentrated silicone on the cuticle edges of the hair whereas the Patent Application shampoo "distributed silicone more evenly."
- X-Ray photoelectron Spectroscopy (XPS) to measure thickness of silicone polymer layer on hair from Si:C:O ratios. This method revealed that the commercial shampoo deposited a significant amount of silicone, and the Patent Application shampoo deposited only one or two molecular layers.
- Instron Ring Compression as a measure of combability

Complex coacervates can also be formed from mixtures of cationic and anionic polymers. This could be the underlying mechanism in shampoos that include a carbomer with cationic polymer to enhance the deposition of silicone and to improve conditioning benefits.[125]

Improving the clarity of conditioning shampoos: Silicones are excellent for adsorbing at interfaces and spreading and wetting on hair to provide smoothing and conditioning benefits but there are challenges in making stable suspensions of silicones in shampoos. Silicones often destabilize foam and usually silicone shampoos are hazy or cloudy. Silicone copolyols were introduced to overcome solubility the silicones in shampoo compositions. However, this drastically reduced the amount of silicone deposited onto hair and compromised conditioning performance. Clear, silicone-containing conditioning shampoos have been formulated by adding trideceth-2 carboxamide MEA to reduce the silicone droplet size sufficiently to maintain a light transmission value of greater than 97%.[126] The desire to formulate transparent conditioning shampoos has led to the need to incorporate the silicone as microemulsified droplets. Such small particles tend to be rinsed away rather than deposited during the shampooing process and coalescence of the droplets can lead to loss of transparency in the product during storage. Attempts have been made to overcome this problem by including silicone emulsions with high internal viscosities; typically greater than 100,000 centistokes. However, the high internal phase viscosity presents the challenge of a deposited material that is difficult to remove and tends to build-up with each consecutive shampooing. Moreover, such build-up can reduce the volume of the desired hair style. It has now been disclosed that shampoo compositions providing superior conditioning to hair while also providing excellent storage stability and optionally high optical transparency or translucency can be obtained by combining a low viscosity microemulsified silicone oil with cationic cellulose polymers and cationic guar polymers having molecular weights of at least about 800,000 and charge densities of at least about 0.1 meq/g.[127] Wells has claimed improved conditioning by including an aminosilicone having a viscosity of from about 1,000 cs to about 1,000,000 cs, and less than 0.5% nitrogen by weight of the aminosilicone in which the aminosilicone has an average particle size of from about 5 to about 50 microns. Conditioning shampoo formulations that include a silicone microemulsion in a conditioning shampoo containing guar hydroxypropyltrimonium chloride and an anionic detersive surfactant have been reported to be clear.[128, 129]

Conditioning shampoos with no cationic conditioning polymer: Conditioning can be achieved by including chain extended silicones in an anionic surfactant-based shampoo. Specific examples of useful silicones include silicone emulsions containing divinyldimethicone/dimethicone copolymer with a viscosity about 1.30×10^8 to 1.5×10^8 mPa.s.[130]

Microscopic identification, with and without dyes, was used to show that the inclusion of polyalkylene oxide alkyl ether particles in a conditioning shampoo led to larger coacervate cohesive flocs (20-500 microns) that resisted shear and gave superior deposition efficiency on hair to confer good wet conditioning.[131]

A conditioning shampoo containing a polyester formed from adipic acid and pentaerythritol provides conditioning for dry hair (possibly from reduced hair friction), with no greasy feel.[132]

Organic conditioning oils such as polybutene have been claimed to confer conditioning benefits from a shampoo.[133] Addition of polybutene is taught to increase the deposition of silicone conditioners, and provides improved conditioning benefits such as wet and dry feel and combing.

O'Lenik disclosed a unique class of alkyl polyglucoside quaternary surfactants with all of the multi-functional attributes of cleansing, conditioning and self preserving.[134] This could have the potential of greatly simplifying the formulation of multifunctional shampoos.

Polymer-particle interaction for deposition of particles on hair: There is a need to increase hair volume and styling from conditioning shampoos, and an interesting method to meet this need is to include particles that can deposit on the hair to increase interfiber friction. It has been revealed that cationic guar polymers can enhance the deposition of particles on hair.[135] Pigment particles can be included for color, friction-conferring particles for style (titanium dioxide, clay, pearlescent mica, silica), and platelet or spherical particles for slip and conditioning (hollow silica, hollow polymer). Shampoos containing ethoxylated alcohols have been found to enhance the deposition of large particle silicones (5 to 2000 microns) and in this case it is claimed that cationic polymer is not required.[136]

Hollow particles can be included in shampoo to increase hair volume in people with fine hair that tends to lie flat.[137,138] Factors that influence hair body and fullness include hair diameter, hair fiber-to-fiber interactions, natural configuration (kinky, straight, wavy), bending stiffness, hair density and hair length.

Deposited hollow particles that can increase fiber-fiber interaction include complexes of gas-encapsulated microspheres such as silica modified ethylene/methacrylate copolymer microspheres and talc-modified ethylene/methacrylate copolymer microspheres; polyesters and inorganic hollow particles.

Cationic polysaccharide polymers having a molecular weight of from about 10,000 to about 10 million and a charge density from about 1.4 meq/gm to about 7.0 meq/gm have been claimed to enhance the deposition of antidandruff particulate ingredients on surfaces.[139] In addition, enhanced deposition of particulate actives, such as zinc pyrithione, (shown on cadaver skin treated in a Franz diffusion cell) has been reported[140] from shampoos comprising a water-soluble silicone (such as silicone quaternium-13, cetyltriethylammonium dimethicone copolyol phthalate, or stearalkonium dimethicone copolyol phthalate), a cationic conditioning agent (such as acrylamidopropyltrimonium chloride/acrylamide copolymer, or guar hydroxypropyltrimonium chloride), a cleansing detergent, and suspending agents (such as Carbomer, hydroxyethylcellulose, and PVM/MA decadiene crosspolymer) to insure homogeneous distribution of the insoluble active.

Visually Attractive Two Phase Systems: There is aesthetic appeal to products that exist as separate phases in the bottle but which mix during application to provide added benefit by interaction of the components of the two previously separate phases. The most obvious way to formulate such products is to use the intrinsic immiscibility of water and oil and early attempts to formulate such systems involved products that contained oil and water layers which were shaken prior to use to produce a metastable emulsion. However, when surfactant is included in the system such a precise phase separation represents considerable difficulty because emulsions are readily formed during manufacturing and packing operations and the aesthetic appearance is lost before it even reaches the bottle. There are known de-emulsifiers and these are widely used in the oil industry. However, for shampoos there is a consumer perception requirement of rapid and voluminous lather and unfortunately, de-emulsifiers and oils are generally effective foam breakers. In 1973, Weiner reported that the inclusion of neutralized polyacrylate as non-emulsifying foam stabilizer yielded phase-separated compositions that overcome the production difficulties and produced phase-separated shampoos that formed a temporary emulsion upon shaking and foamed during use.[141,] Olson described a 2-phase shampoo system that was formed by mixing polar lipophilic shampoo components with non-polar 'lotion' constituents such as mineral oil.[142]

Under appropriate conditions, phase-separated systems can be prepared from polymer solutions or micellar surfactant solutions. If two distinct aqueous phases are desired in a composition, one must take into consideration the thermodynamics of coexisting phases and the driving force for such phase separation comes directly from the chemical thermodynamics of the system: As molecules are constrained by polymerization or by aggregation into micelles, the configurational entropy is reduced and mixing can become unfavorable. Thus, if the free energy of mixing is insufficient, spontaneous phase separation will occur. Phase separation becomes more likely as the micellar aggregates or polymers get bigger. The addition of salts to ionic surfactant micellar systems causes a reduction in the surfactant intra-micellar head-group interaction, and often an increase in hydrophobic interaction. This can cause a pronounced increase in micelle size and consequent phase separation into a surfactant-rich phase and a surfactant-poor phase. This has been achieved by adding mineral salt to induce two distinct layers,[143] and by adding the detergent builder, sodium hexametaphosphate, to cause phase separation. In the former case a thickener is required and the system comprises a surfactant, a thickener, a polyalkylene glycol and a non-chelating mineral salt. The system spontaneously separates into two layers.

A multiphase composition comprising surfactant, betaines, cosurfactant (such as an alkyl ether carboxylate , an acylglutamate; an acylisethionate), and an appropriate concentration of salt forms a stable multiphase system that becomes temporarily uniformly dispensed upon agitation.[144]

Multiphase cleansing products have been introduced that go beyond mere phase separation insofar as the separate phases can be arranged to form visually attractive patterns inside a transparent container.[145,146,147] The phases comprise an aqueous cleansing phase, a benefit phase, and a non-lathering structured phase. The aqueous cleansing phase must be capable of adequate lathering.[148] The benefit phase comprises hydrophobic component(s) or conditioning components.[149] These products are designed at the nanoscale: the structured phase can be a lamellar-phase formed by adding sufficient electrolyte to an appropriate surfactant. [150] Structurants such as starch have been used in personal cleansing formulations[151] but the surfactant itself can be structured. Thus, lamellar phase does exhibit a yield stress that is sufficient to stably suspend the benefit phase. However, the yield stress of lamellar phase can vary dramatically with temperature and in order to overcome this problem the cleansing and benefit phases were density-matched by adding microsphere particles to reduce the specific gravity of the cleansing phase or high density particles to the benefit phase increase its specific gravity. In this context, it is interesting that it has been recently disclosed that controlled phase separation and deposition could conceivably be achieved by loading the desired 'active' phase into hollow-sphere polymer carriers[152] and Wells *et. al.* have reported that certain cationic guar derivatives can enhance the deposition of conditioning additives and/or solid particle benefit agents.[153]

Lamellar phase, especially if it is made from unneutralized long-chain fatty acids, usually displays poor dispersion kinetics and a lather that is slow to build up or slow to rinse off. However, if has surprisingly been discovered that swollen lamellar gels can exhibit both high product viscosity and fast dispersion kinetics if they are formed by combining C16-24 normal monoalkylsulfosuccinates with *n*-alkyl fatty acids of approximately the same chain length.[154] In this context, Guth claimed a composition that was low-irritating to skin and eyes but synergistic in foaming by combining zwitterionic surfactants-fatty acid complexes with sulfosuccinates,[155] and Pratley reported synergistic foaming and mildness from compositions with combinations of specific long-chain surfactants with specific short-chain surfactants and these included fatty acids and alkylsulfosuccinnates.[156] Amine-oxide copolymers have also been claimed as suds-enhancers.[157]

Wipes and skin feel: Two types of cosmetic wipes are currently being marketed: moist wipes which are impregnated with a cleansing or skin-care formulation and dry wipes which have to be moistened before use. Wipes were originally directed towards cleaning and sanitization, but now cosmetic attributes such as skin feel and foam structure are becoming increasingly important. Wipes function to avoid the direct application of cleansing ingredients to the skin and to insure a more regular distribution of the cleansing agent. In the recent years major efforts have been spent to improve the softness and absorbency characteristics of wipe and pad materials to yield wipes of higher absorbency and better softness to alleviate skin irritation upon prolonged or repeated usage.[158] Good foaming, cleansing and skin feel has been reported for wipes impregnated with an emulsifier mixture containing nonionic and amphoteric surfactants, a mixture of wax components containing wax esters, partial glycerides and fatty alcohol ethoxylates and a cationic polymer.[159] The preferred cationic polymer in this case is a cationic guar and this polymer is likely to confer the perceived skin feel benefits. Cationic galactomannans, such as Guar Hydroxypropyltrimonium chloride are commonly used as thickeners deposition aids in skin-care compositions. Most wipes are fragranced, but there is a fraction of consumers who prefer wipes with a barely perceptible hint of fragrance or no perceivable fragrance at all. It is interesting in this context that Hercules has recently reported the development of a cationic guar with reduced odor. The cationic functionality of these polymers is conferred by quaternizing with trimethylamine and trace residuals of this amine can cause the product to have a 'fishy' odor, which may be detectable in products formulated at neutral or basic pH values. Hercules has claimed that a cationic galactomannan with less than 25 ppm of residual trimethylamine has no discernible odor when formulated into personal care compositions.[160]

Silicones for Skin Feel

Skin feel is an important attribute of skin lotions and make-up and, in order to optimize this property, over the last quarter of a century or so, hydrocarbon and emollient oil-in-water emulsions have evolved to silicone-in-water emulsions, to the incorporation of volatile silicones and oils for 'lightness' and eventually to water–in–silicone emulsions. Fluoroalkyl-modified silicones have been proposed to confer oil-repellency on the skin, but these compounds lose some of the 'lightness' sensation that is expected from silicones.[161] The emulsifiers for these products have evolved from conventional hydrocarbon emulsifiers to silicone emulsifiers such as silicone copolyols. Recently it has been revealed that an alternating copolymer of α,ω-organohydrogenpolysiloxane and triglycerol diallyl ether is less drying than silicones and less tacky than glycerol.[161] Certain polyethoxylated silicone polyamides are claimed to be self-emulsifying.[162] Silicone resins give a silky feel to the skin when dried but they go through a tacky, draggy phase during dry down. This tacky stage can be eliminated by using a mixture of amorphous silica, silicone resin (Dimethicone and Trimethylsiloxysilicate) and a low oil absorbing spherical powder.[163]

Silicone resins are typically made by hydrosilylation in which SiH groups are reacted with terminal olefinic groups to form crosslinked siloxane polymers. This limits the range of possible organofunctional groups that may be incorporated into the polymeric structure to create performance advantages in cosmetic formulations. Silicone epoxide chemistry has been developed to overcome these limitations and to increase the diversity of polymeric silicone gels. This chemistry allows the synthesis of polyether-siloxane copolymer crosslinked networks in which certain cosmetic fluids are insoluble but are capable of swelling the network.[164,165] The desired fluids are silicone fluids, hydrocarbon fluids, esters, alcohols, fatty alcohols, glycols and organic oils. The copolymer network gels the fluid to confer the properties of a solid gel material at rest. The composition shows high stability and resistance to syneresis but the fluid may be released from the network by rubbing the composition on the skin to provide the sensory feel characteristic of the silicone fluid. Another approach to forming stable clear gels of volatile silicones is to use siloxane polymers with quadruple hydrogen bonding units.[166] The quadrupole hydrogen bonding units are formed from nitrogen containing compounds that are reacted with isocyanates or thioisocyanates, or that are activated and reacted with primary amines, to obtain a urea moiety that is part of the quadruple hydrogen bonding site on a polysiloxane molecule. These molecules are efficient thickeners of volatile silicones. Since the amount to form gels is tiny, the perceived tackiness on drying is reduced or eradicated.

Silicone resins have also been utilized to confer transfer resistance on color cosmetics such as lipsticks[167] but the early products were perceived by some consumers to give a certain level of discomfort during wear. It has now been revealed that poly(alkyl, hydroxyalkylsiloxanes) are capable of improved color fastness, while conferring satisfactory gloss and satisfactory comfort.

A Skin protectant with good skin feel, and easy application can be obtained by using a block copolymer having a linear polysiloxane-polyoxyalkylene block and an HLB less than 6, in combination with a cosmetic oil, and with a low water content.[168]

Silicone emulsions are known to confer good feel on skin, but these emulsions usually require the use special emulsifiers such as dimethicone copolyol. Non-aqueous emulsions of silicones are useful delivery systems for cosmetic applications, particularly when the presence of water initiates a process that changes the nature of the cosmetic composition.[169] The polyether-siloxane network copolymers can be used to enhance the properties of non-aqueous emulsions, oil-in-water emulsions or water-in-oil emulsions.[164] Clear microemulsions are alleged to be formed spontaneously by simply combining water; a volatile siloxane; a long chain or high molecular weight silicone polyether; and, optionally a cosurfactant such as a monohydroxy alcohol, an organic diol, an organic triol, an organic tetraol, a silicone diol, a silicone triol, a silicone tetraol, and a nonionic organic surfactant.[170]

Glyceryl derivatives of dimethicone are claimed to provide better long-term emulsion stability than silicone copolyol and they spontaneously swell to absorb their own weight of silicone oil.[171]

Silicones can build up with continued use on skin leading to a dull appearance. Esters of hydroxycarboxylic acids and alkyl and alkenyl oligoglycosides are declared to permanently re-emulsify silicones, making them easy to remove after each use and preventing buildup.[172]

Particle–containing emulsions

The emulsifier-free trend in skin lotions and sunscreens has created an interest in particle-stabilized or so-called 'Pickering emulsions. These emulsions are not new; in fact, Pickering produced particle-stabilized emulsions before 1910.[173, 174] What *is* new is the modern ability to tailor-make nanoparticles to specifically stabilize emulsions.[175] Such emulsions are stabilized by a packed layer of particles adsorbed at the oil-water interface. In contrast to conventional surfactant-stabilized emulsions, the particles are less mobile and HLB is not as important. This offers the formulator more latitude and versatility especially in stimuli-responsive compositions that are storage-stable but become unstable and release the oil-phase when applied to the substrate. Polymer particles may be used to prepare Pickering emulsions.[176] Examples of these particles are microfine polyamide particles and microfine polymethacrylates. Stable water-in-oil emulsions are particularly difficult to formulate. It has now been discovered that the inclusion of film-forming polymer can enhance the stability of water-in-oil Pickering emulsions.[176] This is a significant advance for color cosmetics and sunscreens that depend upon film-formers for enhancement of substantivity, waterproofing and transfer-free characteristics. The film formers specified in this invention are standard polymers that have been included in cosmetic compositions for a considerable time. Examples are polyvinylpyrrolidone (PVP), or hydrophobic polyvinylpyrrolidone copolymers (PVP hexadecene copolymer and PVP eicosene copolymer), sodium polystyrenesulphonate, or polyisobutene,.

Currently the broadest cosmetic use of nanoparticles is in sunscreens. Zinc oxide and titanium dioxide and nanotitanium dioxide are effective physical sun blockers. However, if high levels of sun protection factors are required, high levels of these physical blockers must be incorporated into the formulation and this leads to thick emulsion products that are difficult to apply and spread and exhibit a heavy, sticky feel. It has recently been taught that steric stabilization of hydrophobic inorganic particles by telechelic, amphipathic oligomeric block copolymers can confer fluid rheology and while maintaining high levels of inorganic particles in the formulation.[177] The middle block of these triblock copolymers is flexible and rubbery – for example polyisobutylene or hydrogenated polyisobutylene with succinate end groups , or polyisoprene or its hydrogenated derivative with polyethylene oxide end blocks.[178]

In hot, humid climates, consumers use vanishing creams to deliver a high skin friction and a matte finish to overcome the oily skin feel and shiny skin appearance. High friction on skin is perceived as 'less greasy'.[179] In these climates there is also a need to include sunscreens in vanishing creams. One possible way to do this is to include zinc oxide in the composition. However, vanishing creams are emulsified and stabilized with stearic acid/stearates and the zinc oxide particles aggregate in the stearic acid oil phase and at the relativity high processing temperature (80°C) they react to form zinc stearate which is a solid that detracts from the "feel" and "rub-in" qualities of the vanishing cream. Pigments can be rendered hydrophobic by treatment with dimethicone/ methicone and this usually improves the dispersion of inorganic pigments in the oil phase. However, hydrophobic zinc-oxide reacts even more readily to form zinc stearate and the resulting aggregates require considerable mixing energy for dispersion in the composition. Moreover, after cessation of mixing the particles can re-aggregate to produce an unstable, inhomogeneous product. It has now been discovered that if the temperature is closely controlled below 80°C, a zinc stearate shell is formed on zinc oxide nanoparticles and this results in disaggregation, easier mixing and more stable vanishing creams containing zinc oxide sunscreen.

Porous microparticles can be used to deliver active agents directly into the pilosebaceous ducts.[180] In order to enter the pilosebaceous unit these particles must have a volume-average diameter of less than 10 micrometers and they must have a BET surface area in the range of 2.5 square meters per g. to 1000 square meters per g. The BET[181] (Brunauer-Emmet-Teller) method measures the total surface area including that inside the pores and it is the method used for the international standard ISO 5794/1 (annex D). The particles are forced into the pilosebaceous unit by massage. The active ingredients that can be delivered include antibacterial agents, antifungal agents, sebum regulators, sebum stimulators, keratolytic agents, acne-treatment agents, antibiotics, hair loss inhibitors and also hair growth stimulators, hair growth inhibiting agents,anti-dandruff agents, antioxidants, astringents and pore-reducing agents, antiperspirant agents,vitamins, and anti-inflammatory agents.[180]

Steric stabilization by block copolymers can also be used in solid water-in-oil emulsion forms (makeup foundation) to confer slipperiness and easy disintegration when the products are taken up by the fingers or a sponge. These stabilized products show a desirable shininess during disintegration. In this case the block copolymers are exemplified by PEG-45/ Dodecyl Glycol Copolymer[182] Amphiphilic polymers may be used to prepare mini-emulsions that are essentially free of conventional surfactant and it has been disclosed[183] that a commercially available styrene/ acrylic copolymer is appropriate for this purpose and a reactive hydrophobic comonomer may be combined with the amphiphilic polymer in order to confer mini-emulsion stability.

Stimuli-Responsive Polymeric Systems

Stimuli-responsive polymers are materials that exhibit a particular desired property that is triggered to change suddenly by a specific change (*e.g.* temperature, pH, ionic strength) in its physical environment. Such materials are of interest to cosmetic chemists because they can aid in processing and manufacture and they allow the formulation of products that exhibit desired attributes in the storage container and a tunable change of properties when added to the substrate.

Thermally–responsive polymers: Thermally reversible viscosifying polymers exhibit a dramatic increase in solution viscosity when they are heated above a threshold temperature. thermal gelling was observed for aqueous compositions containing polymers having a poly (2-acrylamido-2-methylpropanesulphonic acid) backbone and grafted chains of polyoxyethylene and polyoxypropylene or polyoxybutylene.[184] These polymers can be compositionally 'tuned' to be fluid at room temperature but to gel at skin temperature[184] upon traversing an effective lower critical solution temperature that induces micellization of the amphipathic polymers.[185, 186] Hoffmann introduced the concept of critical micelle temperature to describe the dilute solution property of these molecules to aggregate into micelles above a certain critical temperature.[185]

Gelation in response to a temperature stimulus can also be useful as a processing aid. For example, for the case of preparing multiple emulsions. The normal procedure for preparing water/ oil /water multiple emulsions is to prepare the internal emulsion separately from the external base and then to mix the phases together gently to avoid escape of the internal water-phase into the external aqueous phase. Processing difficulties are encountered when the viscosity of the oil-phase is significantly higher than that of either of the aqueous phases. Mercier has reported that the use of amphipathic reverse thermal gelling polymers can raise the viscosity of both the internal and external aqueous phases and thereby enable one pot mixing of multiple phase emulsions without breaking the W/O droplets and thereby mixing of the internal aqueous phase and the external aqueous phase is avoided.[187] Upon cooling below the critical micelle temperature, such lotions become stable low viscosity multiple emulsions. Examples of reverse thermal gelling amphiphilic block polymers that are suitable for use in this application are PEG-30 Dipolyhydroxystearate, Poloxamer, and polymers containing polyalkylpolyether-grafted *poly*-dimethylsiloxane blocks.

One class of such polymers is prepared by end-capping Poloxamers with poly(acrylic acid).[188] These polymers enable the preparation of multiple emulsions.[189] Water-in-oil-in-water (W/O/W) emulsions are systems in which small water droplets are dispersed inside larger oil droplets that are in turn dispersed in an aqueous continuous phase. They have been proposed as suitable formulation vehicles for 2-component actives, or triggered actives in which the chemical properties of the internal aqueous phase and external aqueous phase are different. The objective of the ideal WOW formulation is stability in the container and then mixing of the internal and external phases to trigger the desired response when the emulsion is applied to the skin (or other desired substrate). Formulation of these emulsions is not trivial and their commercialization has been retarded by processing and stability constraints.

Two mechanisms have been proposed for mixing of the actives in the internal and external aqueous phases:

- diffusion of the active agent through the oily barrier to the other aqueous phase
- droplet-bursting either by dilution in a hypo-osmotic solution or by the application of a shear stress.

For droplet-bursting under shear, the droplets disrupt when the shear stress exceeds the droplets'cohesion stress. The cohesion stress is given a capillary number $Ca = \eta_c Gr/\sigma$, where η_c is the continuous phase viscosity, G the shear rate, r the radius of the globule at rest and σ the interfacial tension between oil and water. Bursting occurs when this capillary number exceeds a critical value Ca_{cr} close to unity. Thus bursting is favored by a decrease in the oil/water interfacial tension or increases in shear rate, droplet size and continuous phase viscosity. The continuous phase viscosity can be easily increased with conventional thickeners to enhance probability of the droplet disruption. However, while droplet disruption is desirable on skin, it is not desirable during processing. There is a need, therefore, for thickeners that do not viscosify during processing but which do thicken under the conditions of application to skin. The thermally reversible viscosifying polymers described above can be designed to thicken at skin temperature but to give lower continuous phase viscosity at room temperature. Thus. if the W/O/W emulsions are processed at room temperature, droplet disruption can be minimized; while the triggered increase in viscosity upon application to skin will favor mixing of the internal and external aqueous phases.

Side-chain crystalline polymers are clamed to thicken a wide-range of oils.[190] These polymer comprise a water soluble (or polar) backbone with crystallizable long chains arranged along the polymer like teeth on a fine comb. The melting temperature can be tuned the distribution of side-chain lengths and their content along the chain. These polymers are dispersed in the desired oil and heated above their melting point. Thickening occurs upon cooling below the melting point of the side-chains. These interesting stimuli-responsive copolymers have been used in agriculture to coat seeds. The coating is hydrophobic and protects the seed if the side chains are crystallized. When the side chains melt, water intrusion into the seed occurs and the seed germinates. By adjusting the coating to certain temperatures the seeds can be planted early in spring or even in winter where they remain ready to grow when the ground reaches the triggering temperature.

<u>*Self-heating compositions*</u>: Heat producing cosmetic compositions produce pleasant sensations when exposed to skin. Shaving creams, hand lotions, body lotions, lubricants, and facial preparations, including masks and depilatories are all products that would benefit if they could spontaneously warm upon

application to the skin. With this in mind, there have been regular attempts to introduce self heating compositions. Self-warming lubricating gels can use the heat of solution generated from moisture on the target substrate.[191,192] The warming compositions contain polyols preferably PEG-300 or PEG-400 and these rise in temperature by as much as 5°C when diluted by moisture on the substrate surface. T he compositions also contain thermal insulators.

Methyl salicylate provides a sensation of warming but this compound is irritant to skin. Heating can be caused by increasing blood circulation. This is the mechanism of L- Arginine.[193] Generally, the warm sensation is generated from heat of hydration of the product with water on the skin's surface and the products are necessarily anhydrous. Attempts have been made with finely divided solid absorbent materials such as silica gel, activated alumina and synthetic zeolites[194, 195]. However, the amount of heat generation was small for compositions containing reasonable concentrations of these powders. Gott introduced iron oxide redox systems in order to generate sufficient heat.[196]. This self-warming composition comprised a silicone oil or carboxylic ester as the skin conditioning agent; and a redox system based upon iron powder and a high surface area charcoal catalyst . A two-stage chemically heated liquid soap composition depended upon a novel double reductant and single oxidant redox system of a hydrogen peroxide and a combination of sodium sulfite and ascorbic acid with a suitable catalyst.[197] Boron compounds such as triethoxyboroxine reacts exothermically with water and this has been used in self-heating shaving creams.[198] Polyols can heat by hydration upon topical application and they are claimed to be non- irritating to mucous membranes.[199]. It has now been revealed that self-warming compositions can be based upon the heat of hydration of polyvinylamine or polyethyleneimine.[200] An anhydrous silicone lotion containing 30 percent polyethyleneimine produced a warm sensation a few seconds after topical application. One would expect the skin to be impermeable to polymers, especially polymers with a strong enthalpic interaction with stratum corneum.

Stimuli responsive polymers in decorative cosmetics: Decorative cosmetics often depend upon film-formers to give the product durability on the skin. Volatile solvents are included in these formulations, to allow dissolution of the film-former in the product and good film formation on the skin upon evaporation of the solvent. However, as the polymeric film is formed in this mode, the applied film can shrink and penetrate wrinkles and blemishes on the skin surface. This opposes the objective of applying foundation to camouflage surface blemishes on the skin.

Ferrari and Tournilhac have attempted to overcome this drawback by depositing polymers that increase in volume as the composition dries on the surface of skin.[201] They exploit the fact that polymer molecules adopt an expanded conformation when they are dissolved in a good solvent and a collapsed conformation in a poor solvent. By formulating polymers in a mixture of a nonvolatile good solvent and a volatile poor solvent, they disclose that they are able to form deposits that increase in volume with time and keep the skin defects camouflaged for a long time. The compositions are exemplified by poly (methylstyrene–*co*–2–ethylhexyl acrylate), and poly (styrene–*co*–2–ethylhexyl acrylate/isobutyl methacrylate) and a mixture of kaolin and of poly-(alkylstyrene) and (poly(alkylstyrene) in the good, nonvolatile solvent isopropyl myristate and the poor, volatile solvent isododecane.

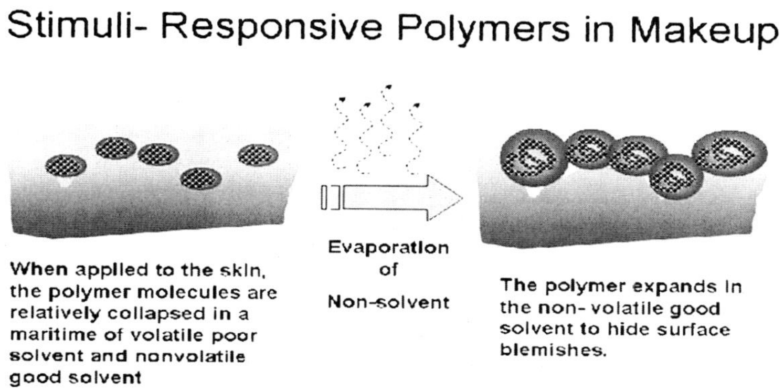

Figure 5: Polymer expansion in a good solvent is one mechanism to camouflage blemishes and wrinkles on skin. (Reproduced from HAPPI *April 2006. Copyright 2006 R. Y. Lockhead.)*

Antiseptic Compositions

Gelled alcohol hand sanitizers are now commonplace. In early versions, alcohol solutions were thickened with hydroxypropylcellulose and the thickener increased the contact time of the antimicrobial alcohol by retarding the evaporation of the alcohol.[202]. Later versions[203] were formulated with alcohol and a cationic antibacterial; they included a silicone wax for moisturizing properties and they were thickened using a Carbomer for better skin feel characteristics. The inclusion of aromatic alcohols and quaternary antimicrobial compounds enhance the broad spectrum effectiveness of these products.[204] Polymeric antimicrobials are now being developed and there is a possibility that these could be included in hand sanitizers for more durable antimicrobial effectiveness.

Polymeric Antimicrobials

In recent years there has been a consumer-driven trend, especially in Europe, to reduce the amount of preservations in cosmetics. No ethical manufacturer would inflict an inadequately preserved product on the consumer. Manufacturers are, therefore, interested in exploring alternative routes to preservation. Short peptides having antimicrobial activity have been discovered and are targeted towards reducing the microbial activity on or in skin with respect to conditions such as acne.[205] Polypeptides that contain sequences that are rich in amino acids with positively charged side groups enhance cell penetrance and it is believed that penetration of the cell membrane is an essential step in the mechanism involved in the antibacterial functioning of these molecules. [205, 206] Polyhexamethylenebiguanide[207] , polyornithine, and polylysine are two candidate polypeptides. Polypeptides may represent an acceptable approach to preservation of personal care products and it is interesting. In this context, that epsilon polylysine is being advanced as a 'safe' water-soluble preservative for skin-care compositions.[208]

UV Absorbing Polymers

UV absorbing water soluble polymers have been claimed to protect substrates from the effect of ultraviolet light wavelength of 200 to 420 nm.[209] The advantage of these polymers is that they allow delivery of a UV-absorber directly from an aqueous solution from a polymeric form that bodes well for deposition on the desired substrate. The polymers combine four monomer types: the first is a vinyl or acrylic monomer that absorbs ultraviolet light radiation of wavelength in the range 200 to 420 nm; and the other three monomers are free-radical polymerization water-soluble monomers. The UV absorbing monomers are selected from N-[3-(dimethylamino)propyl]methacrylamide-N-(3-bromopropyl)phthal-imide quarter-nary salt (DMAPMA-PQ), N-[3-(dimethylamino)propyl]-methacrylamide 1-chloromethylnaphthalene quaternary salt (DMAPMA-MNQ), (3-allyloxy-2-hydroxypropyl)-[3-(2-hydroxybenzoyl-amino)propyl]-dimethyl ammonium hy-droxide, [(4-carboxy-3-hydroxyphenyl-carbamoyl)methyl]-dimethyl-[3-(2-methyl-acryloylamino)propyl] ammonium hy-droxide, and 4-methacrylamidosalicylic acid (4-MASA).

Summary

Polymeric materials are critically important to the creation of modern cosmetics and personal-care products. Low VOC regulations have driven the polymer manufacturers to create fixatives that can be delivered from aqueous solutions . In this context, block-copolymers and microgel thickeners that also form films have been developed. Hydrophobically-modified cross-linked polyelectrolytes have been developed as polymeric emulsifiers for the triggered release of the oil phase when application to the substrate – this has found use in sports sunscreens. Skin emulsions are trending towards lower pH values and ingredients such as alpha hydroxyacids are being included. Hydrophobically-modified Acryloyldimethyltaurate copolymers have been developed thicken such systems in this relatively hostile environment for conventional rheology-modifiers. Associative thickeners are being extensively used in hair-coloring products to gel the product on the substrate, but to simultaneously allow good color development. Free radical, living polymerization techniques hold promise for the development of systems having well-defined mesomorphic morphology for the design of novel rheology-modifiers and delivery systems. A horde of new cationic polymers have been developed for conditioning purposes, but conditioning shampoos continue to be dominated by the workhorses, Polyquaternium-10 and Guar hydroxypropyltrimonium chloride.

Stimuli-responsive polymers are being used in applications from make-up that camouflages wrinkles to the facile processing of multiple emulsions, to thermally-responsive systems that respond to the surface of skin.

References

1. Gottschalk, T.E.; Mc Ewen ,G. N: *International Cosmetic Ingredient Dictionary and Handbook, Eleventh Edition*, **2006**, The Cosmetic, Toiletry and Fragrance Association.
2. Dubief, C; Giroud F.; and Rolla, I.; *US Pat Application 2004/00338246*, Feb 19, **2004**.
3. Dupuis, C.; *US Pat 6,592,854*, (**Jul 15, 2003**), assigned to L'Oréal
4. Kim, S.N.; Sanner, A.; and Schehlman, V: *US Pat 6,579,517,* (**Jun 17, 2003**) assigned to BASF
5. Kim, S.N.; Sanner, A.; Hossel, P.; *US Pat 6,800,276*, (**Oct 5, 2004**), assigned to BASF
6. Matyjaszewski, K.; Coca, S.; Gaynor, S.; Greszta, D.; Patten , T.; Wang, J.; Xia, J.; *PCT Int. Appl. WO 9718247 A1*, (**1997**),
7. Matyjasezwski, K:, *J. American Chem Soc,* **1995**, *117*, 5614
8. Kobatake, S.; Harwood, H.J.; Quirk, R.P.; *Macromolecules,* **1997**, *30* 4238.
9. Nesvadba, P. ; Kramer, A.; Steinmann, A.; ; Stauffer, W. ; **1999**, PCT Int. Appl . WO 9903894 A1.
10. Malmstroem, E.; Hawker, C.; Craig, J.; *Macromol Chem Phys,* **1998,** *199,* 923-935
11. Malmstroem , E.; Hawker, C. ; Craig, J. ; **1998**, PCT Int. WO-A-98/58974
12. Chong, Y. K.; Le Tam P. T.; Moad, G.; Rizzardo E.; Thang , S.H.; *Macromolecules,* **1999**, *32,* 2071-2074.
13. Mougin, N.; *US Pat 6,805,872*, **Oct 19, 2004**, assigned to L'Oréal.
14. Vitale, M .; Tolchinsky, M.; Martino, G.; Solarek, D.; Cottrell, I.; *US Pat 6,562,325*, **May 13, 2003**, assigned to National Starch.
15. Asaoka, S.; Koyama, K.; Tsuzuki, T.; Hashimoto, T.; *US Pat 6,737,069,* **May 18, 2004**

16. Morschhauser, R; Kayser, C; Loffler, M.; *US Pat 6,833,419*, **Dec 21, 2004**, assigned to Clariant.
17. Green, M; Dungworth, H. R.; Gavin, D. E.; Ridley, D.B.; *US Pat 6,833,406,* **Dec 21, 2004**, assigned to Ciba
18. Castaneda, J. Y.; Lochhead, R. Y.; European Patent 482417 A1; April 29, 1992; assigned to BF Goodrich.
19. Lochhead, R. Y. ; Castaneda, J. Y. and W. J. Hemker , W. J.; European Patent 268164 A2, May 25, 1988; U.S. Patent 5,004,598, April 2, 1991; assigned to BF Goodrich.
20. Thibeault, J.C.; Sperry. P.R.; Schaller, E.J.; ACS Symposium Series 213, J.E. Glass editor, **1986**, Ch 20, *375.*
21. Morschhauser; Roman (Mainz, DE); Loffler; Matthias: United States Patent 6,645,476; November 11, 2003; assigned to Clariant GmbH.
22. Morschhauser; Roman; Kayser; Christoph; Loffler; Matthias; United States Patent 6,833,419; December 21, 2004, assigned to Clariant GmbH.
23. Deckner, George; United States Patent Application 20050031568, February 10, 2005; assigned to the Procter & Gamble Company.
24. Deckner, George; United States Patent Application 20050031569, February 10, 2005; assigned to the Procter & Gamble Company.
25. Deckner, George; United States Patent Application 20050031660, February 10, 2005; assigned to the Procter & Gamble Company.
26. Deckner, George; United States Patent Application 20050032916, February 10, 2005; assigned to the Procter & Gamble Company.
27. Cardinali, M.S.; Lam, T.Y. Presented at PCIA Manila, 2003
28. L'Alloret, F.; Aubrun-Sonneville, O; Simonnet; J-T.; *United States Patent 6,998,426,* **February 14, 2006**; assigned to L'Oreal.
29. L'Alloret, F.; Aubrun-Sonneville, O; Simonnet; J-T.; *United States Patent Application* 20060030655; **February 9, 2006**; assigned to L'Oreal.
30. Williams, L.: *United States Patent 5,422,112,* **June 6, 1995**; assigned to Chesebrough-Pond's USA Co., Division of Conopco, Inc.
31. Deckner, G. E.; Lombardo, B. S.; *United States Patent 5,874,095;* **February 23, 1999**; assigned to Richardson-Vicks
32. Ito, K.; Mori, Y.; *United States Patent Application 20060046949*; **March 2, 2006**
33. Lorant, R.; *United States Patent 5,952,395,* **September 14, 1999**; assigned to L'Oreal
34. Shinoda, K.; Friberg, S. *Emulsions and Solubilization*; **1986**, Wiley, New York,
35. Sebillote-Arnaud, L.; Lorant, R.; *United States Patent 5,891,452*; **April 6, 1999**; assigned to L'Oreal
36. Suares, A. J., Zhang, J. H., *United States Patent 6,986,895*, **January 17, 2006**, assigned to Unilever Home & Personal Care USA.
37. Morschhauser, R.; Löffler, M.; Maier, I.; *United States Patent 6,964,995*; **November 15, 2005**; assigned to Clariant .

38. Morschhäuser, R.; Glauder, J.; Löffler, M. ; Kayser, C. ; Tardi, A.; *United States Patent 6,891,011,* **May 10, 2005**, assigned to Clariant
39. Lai, J.; Hsu, S.-J. R.; Tamareselvy, K.; *United States Patent Application 20060039939;* **February 23, 2006**.
40. Raether; R. B.; Brinkmann-Rengel, S.; Haremza; S.; *United States Patent 7,008,990;* **March 7, 2006**; assigned to BASF
41. Hawker, C. J.; *Acc. Chem. Res.;* **1997**, *30(9)*; 373;
42. Wang, J. S.; Matyjaszewski, K. *J. Am. Chem. Soc.*, *117*, 5614, (1995)
43. Thomas, D.B. ; Sumerlin, B.S., Lowe, AB; McCormick, C.L.; *Macromolecules*, 36, 1436., (2003).
44. Convertine, A. J.; Sumerlin, B.S.; Thomas, D.B.; Lowe, A. B; McCormick, C.L.; *Macromolecules*, 36, 4679, (2003).
45. Cameron, A N.S.; Corbierre, M.K. and Eisenberg, A.; *Can. J. Chem.,* **1999**, *77*, 1311.
46. Xu, R.L.; Winnik, M.A.; Hallett, F.R.; Riess, G.; Croucher, M.D.; *Macromolecules*, **1991**, 24, 87-93.
47. Wilhelm, M.; Zhao, C.L.; Wang, Y.; Xu, R.L.; Winnik, M.A.; Mura, J.L. Riess, G.; Croucher, M.D; *Macromolecules*, **1991**, 24, 1033-1044.
48. Arien, A. M. E.; Brewster, M.E.; Nathan, A.; Rosenblatt, J.; Ould-Ouali, L. M.; Preat, V.; *United States Patent Application 20060034797*; **February 16, 2006**.
49. Simonnet, J.-T.; *United States Patent Application 20060024337*; **February 2, 2006**.
50. Yekta, A.; Duhamel, J.; Adiwidjaja, H.; Brochard, P.; Winnik, M.A.; *Langmuir*, **1993**, *9*, 881-883.
51. Liu, T; Burger, C.; Chu, Br; "*Nanofabrication in polymer matrices*", Prog. Polym Sci., 28, 5, (2003).
52. Ma, Q.; Wooley, K. L.; *J. Polymer Sci. Part A: Polymer Chemistry,* **2000**, *38*, 4805.
53. Cao, T.; Munk, P; Ramireddy, C.; Tuzar,Z.; Weber, S.E.; '*Fluorescence Studies of amphiphilic poly methacrylic acid)- block-polystyrene –block-poly(methacrylic acid) micelles',* Macromolecules (1991), **24**, 6300.
54. Cadena, N.; Destarac, M.; Herve, P.; Wilczewska, A. Z.; United States Patent Application 20050180937; August 18, 2005; Rhodia Inc.
55. Cadena, N.; Destarac, M.; Herve, P.; Wilczewska, A. Z.; United States Patent 6,906,128; June 14, 2005; assigned to Rhodia Chimie.
56. Yeung, D. W-K.; Bergeron, V.; Bodet, J. F.; Sivik, M. R.; Kluesener, B. W.; Scheper, W. M.; United States Patent Application 20050113272; May 26, 2005; assigned to Rhodia Inc.
57. L'Alloret; F.; "*Composition for cosmetic or dermatological use containing a triblock polymer",* United States Patent 6,994,846, February 7, 2006, assigned to L'Oreal.

58. Lion; Bertrand; *Hyperbranched copolymer comprising monomers of choice, a composition, and a cosmetic method;* United States Patent Application 20060030686; February 9, 2006
59. Marschner; Frank W.; Kenkare; Divaker B; Bowers; James H.; *Anhydrous antiperspirant composition* United States Patent 4,526,780; July 2, 1985; assigned to Colgate-Palmolive
60. Tanner; Paul R; Nunn, Jr.; Randolph G.; Luebbe; John P.: *Low residue antiperspirant creams,* United States Patent 5,019,375; May 28, 1991; assigned to Procter & Gamble.
61. Shin; Chung T; *Anhydrous semi-solid antiperspirant suspension*; United States Patent Application 4,937,069; June 26, 1990; assigned to Bristol-Myers Squibb.
62. Orr; Thomas V.; *Antiperspirant creams;* United States Patent 5,069,897; December 3, 1991; assigned to Procter & Gamble
63. Abrutyn; Eric S.; Bahr; Bradley C.; Legrow; Gary E.; Schulz, Jr.; William J.; *Underarm formulations containing alkylmethylsiloxanes;* United States Patent 5,225,188, July 6, 1993, assigned to Dow Corning.
64. Putnam; Christine Marie; Bretzler, Eric John; Swaile; David Frederick; *Antiperspirant cream compositions comprising select triglyceride gellants;* United States Patent 5,718,890; February 17, 1998; assigned to Procter & Gamble.
65. Stepniewski; George J.; Konik; Richard A.; Dreher; John D.; Cioca; Gheorghe; Cohen; Isaac D.; *Stable water-in-oil emulsion system,* United States Patent 5,599,533; February 4, 1997, assigned to Estee Lauder
66. Healy, Lin Lu; Cunningham, Jack C. JR.; Morrisson, David S.; Song, Wei; Butuc, Gina; United States Patent Application 20050004274, January 6, 2005.
67. Guenin; Eric; Mattai; Jairajh; Afflitto; John; Hogan; John; Jonas; John; Lee; Wilson; Linn; Elizabeth; Munsayac; Rosemary; Tang; Xiaozhong; Potechin; Kathy; *Effective soft solid personal care product;* United States Patent 7,011,822; March 14, 2006, assigned to Colgate-Palmolive.
68. Trigg, David Leigh; Chen, Yin-Chang;; Chen, Minghua; United States Patent Application 20050013784, January 20, 2005, assigned to the Procter & Gamble Company.
69. Zolotarsky, Yelena; Lamont Sims, Cary; Botschka, Ellen S.; Krautsieder, Barbara; United States Patent Application 20050019291, January 27, 2005; filed by International Specialty Products.
70. Yang, J.; *US Pat App 2004/0255399*, **Dec 23, 2004**.
71. Casperson, S.; Pohl, S.; Wong, M.; *US Pat 5,376,146*, **Dec 27, 1994**, assigned to Bristol-Myers Squibb
72. Lutz, C. W. ; Cohen, L. E.; *US Pat 4,130,501*, **Dec 19, 1978**, assigned to FMC

73. De la Mettrie, R.; Boudy, F.; *US Pat 6,010,541*, **Jan 4, 2000**, assigned to L'Oréal
74. Audousset, M-P.; *US Pat App 2005/0000037*, **Jan 6, 2005**.
75. Pohl, S.; Hnatchenko, M.; *US Pat RE 33,786*, **Jan 7, 1992**, assigned to Clairol.
76. Landoll, L.M.; Sau, A.C.; Synthesis, *ACS Adv Chem Ser 223,* **1989**, 343-364.
77. Hoffmann, H.; Wanka, G.; *Colloid Polym Sci,* **1990**, *268*, 101.
78. Xu, B.; *Langmuir,* **1997**, 13 6896; Xu, B.; *Langmuir,* **1997**, *13* 6903.
79. Yekta, R.; Winnik, M.; *Macromolecules,* **1995**, *28*, 956.
80. Annable, T.; Buscall, R.; *J Rheol,* **1993**, *37,* 695.
81. Glancy, C.W.; Steinmetz, A.L.; *US Pat 5,914,373*, **Jun 22, 1999**, assigned to United Catalysts, Inc.
82. Legrand, F.; Allard, D.; *US Pat 6,800,096*, **Oct 5, 2004**, assigned to L'Oréal.
83. Allard, D.; Legrand, F.; *US Pat 6,800,098,* **Oct 5, 2004**, assigned to L'Oréal
84. Wong, M.; Memisha, E.; *US Pat 6,805,856*, Oct 19, 2004, assigned to L'Oréal.
85. Berthiaume, M.D.; *J Soc Cos Chem,* **1995**, 46, 231.
86. Devin-Baudoin, P.; Sabbagh, A.; *US Pat 6,824,764*, **Nov 30, 2004**, assigned to L'Oréal.
87. Butts, M.D.; Nye, S.A.; Byrne, C.M.; Katritzky, A.R.; Merkert, J.W.; *US Pat 6,841,164,* **Jan 11, 2005**, *US Pat 6,844,002,* **Jan 18, 2005**, assigned to General Electric Company.
88. Plos, G.; Samain, H.; *US Pat Appl 2004/0258641*, **Dec 23, 2004**
89. Khan, G.F.; Guskey, S.M.; Wells, R.L.; **Jan 6, 2005**, *US Pat Application 20050002892.*
90. Dieing, R.; Hossel, P.; Kothrade, S.; Sanner, A.; Zeitz, K.; Raubenheimer, H-J.; Schehlmann, V.; *US Pat 6,682,725*, **Jan 27, 2004** , assigned to BASF.
91. Zofchak, A.; Obeji, J.; *US Pat 6,800,716*, **Oct 5, 2004**, assigned to Alzo International.
92. K Ivanova and SK Pratley, US Pat Application 20050002871, **Jan 6, 2005**.
93. *Patent Application GB 0102657.4,*(**filed Feb 2, 2001**)
94. Samain, H.; Livoreil, A.; *US Pat Application 20040265256*, **Dec 30, 2004**.
95. Roreger; Michael; Schnitzler; Iris; Li; Qian-Yi; *Soap preparation with air bubbles;* United States Patent Application 20060052263; March 9, 2006; assigned to SCS *Skin* Care Systems
96. Kasturi; Chandrika; Schafer; Michael Gayle; Spears; Marsha Jean; Hutton, III; Howard David; Sivik; Mark Robert; Kluesener; Bernard William; Scheper; William Michael; United States Patent 6,827,795, December 7, 2004, assigned to Procter & Gamble

97. Sivik, Mark Robert; Bodet, Jean-Francois; Kluesener, Bernard William; Scheper, William Michael; Yeung, Dominic Wai-Kwing; Bergeron, Vance; United States Patent Application 20050026803, February 3, 2005, assigned to the Procter & Gamble Company.
98. Aronson, A.S.; Bergeron, V., Fagan, M.E.; Radke, C.J.; *Colloid Surf. A*, **1994**, *83*, 109
99. Bergeron, V.; Langevin, D; Asnacios, A.; *Langmuir*; **1996**, *12*, 1550.
100. Geary; Nicholas William; Coffindaffer; Timothy Woodrow; Brown; Mark Anthony; El-Nokaly; Magda; United States Patent 6,849,584; February 1, 2005, assigned to the Procter & Gamble Company.
101. Deckner, George E; Manchuso, Scott E.; Monsueir, W. J; Rodriguez, W.R.; Sine, Mark R; *Rinsable skin conditioning compositions,* United States Patent 6,699,688. , March 2,2004; assigned to Procter & Gamble.
102. Knopf, Michael A; Polk, Michele; Lucia, Frank A. III; Wohland, William C.; Macchio, Ralph; *Deodorant body wash with lotion;* United States Patent Application 20060008434; January 12, 2006
103. Knopf, Michael A; Polk, Michele; Lucia, Frank A. III; Wohland, William C.; Macchio, Ralph; *Cleansing foaming sunscreen lotion;* United States Patent Application 20050265936; December 1, 2005
104. Wei, Karl Shiqing; Thomas, Cheyne Pohlman; Taylor, Rebecca Ann; Tanner, Paul Robert; Stella, Qing; Smith, Edward Dewey III; Clapp, Mannie Lee; *Multi-phase personal care composition;* United Stint's Patent Application 20050100570; May 12, 2005; assigned to Procter & Gamble.
105. Aronson, Michael Paul; Tsaur, Liang Sheng; *Method of enhanced moisture or reduced drying using wet-skin treatment compositions;* United States Patent Application 20050244356; November 3, 2005; assigned to Unilever Home & Personal Care USA Division of Conopco.
106. Clapp, Mannie Lee; Taylor, Rebecca Ann; *Wet skin treatment compositions comprising gel-networks;* United States Patent Application 20040223992; November 11, 2004; assigned to Procter & Gamble.
107. Bonnardel, Valeria; Catterson, Florence; Gestman, Anja; *Emulsifier,* United States Patent Application 20050208009, September 22, 2005.
108. Jorgensen, Lise W; McManus, Richard L.; Spengler, Eric G, *Skin Moisturizer Lotions containing a sebum control agent,* United States Patent Application 20020096341, July 18, 2002.
109. Goddard, E.D. and Ananthapadmanabhan, K.P., eds.; *Interactions of Surfactants with Polymers and Proteins*; CRC Press, Florida; Chapter 4; 1993.
110. Lepilleur, C. A.; Fruscella, J. A.; United States Patent Application 20050129643; June 16, 2005; Noveon IP Holdings Corp.
111. Bolich, R. et.al; U.S. Patents 5,104,606 and 5,106,609 , Assignee :P&G
112. Cadena, N.; Destarac, M.; Herve, P.; Wilczewska, A. Z.; United States Patent Application 20050180937; August 18, 2005; Rhodia Inc.

113. Cadena, N.; Destarac, M.; Herve, P.; Wilczewska, A. Z.; United States Patent 6,906,128; June 14, 2005; assigned to Rhodia Chimie.
114. Yeung, D. W-K.; Bergeron, V.; Bodet, J. F.; Sivik, M. R.; Kluesener, B. W.; Scheper, W. M.; United States Patent Application 20050113272; May 26, 2005; assigned to Rhodia Inc.
115. O'Brien, M. J.; United States Patent Application 20050197477; September 8, 2005; GEAM - SILICONES – OSI .
116. Kuwata, S.; Sakuta, K.; Mori, S.; United States Patent *4,987,169;* January 22, 1991; assigned to Shin-Etsu Chemical Co., Ltd.
117. Biggs, T. N.; LeGrow, G. E.; United States Patent *5,493,041;* February 20, 1996; assigned to Dow Corning Corporation
118. Kilgour, J. A.; Powell, V, Van Valkenburgh; United States Patent *5,760,116;* June 2, 1998; assigned to General Electric Company
119. Kilgour, J. A.; Chaiyawat, A; Kuo-Tsai, G. L.; United States Patent Application 20050165198; July 28, 2005; GE Plastics.
120. Parran, J.; United States Patent 3,761,418; September 25, 1973; assigned to the Procter & Gamble Company.
121. Fair, M. J.; Massaro, M.; Crookham, H.; Rattinger, G. B.; Dalton, J. J.; Farrell; T. J.; Shafer. G.; United States Patent *6,057,275;* May 2, 2000; assigned to Unilever Home & Personal Care USA, Division of Conopco, Inc.
122. Niemiec; S. M.; Yeh, H.; Gallagher, R.; Ho, K. L.; United States Patent 6,908,889; June 21, 2005; assigned to Johnson & Johnson Consumer Companies, Inc.
123. Shah, P.; Sivakumar, A.; United States Patent Application 20050159330; July 21, 2005; assigned to Unilever Home & *Personal Care* USA, Division of Conopco, Inc.
124. Niemec S. M.; Yeh, H.; Gallagher, R.; Hoe, K.L.; U.S. Patent application 20030176303Al , September 18, 2003; filed by Johnson & Johnson.
125. Maubru, M.; Liebard, B., *United States Patent Application 20030108503,* **June 12, 2003**
126. Denzer, H.; Abe, H.; Pytlik, M.; Jansen, R.; Buhmann, A.; *United States Patent 6,803,050.* **October 12, 2004**; assigned to Kao
127. Peffly, M.M.; Hilvert, J.E.; *United States Patent Application 20050158266*; **July 21, 2005**; assigned to The Procter & Gamble Company.
128. Peffly, M.M.; Chang, D.W.; *United States Patent Application 20040234483*, **November 25, 2004**; assigned to P&G
129. Peffly, M.M.; Chang, D.W.; *United States Patent Application 20040234484,* **November 25, 2004**; assigned to P&G
130. Stella, Q.; *United States Patent Application 2003014317,* **July 31, 2003.**
131. Royce. D.R.; Wells, R.,; *U.S. Patent application 2003014374 Al*, **July 31, 2003**; Assignee: Procter & Gamble

132. Royce. D. R.; Wells, R.; *U.S. Patent application 20030138392*, **July 24, 2003**. Assignee: Procter & Gamble
133. Guskey; S. M.; Coffindaffer; T. W.; Schrader; E. M.; Wells; R. L; Baravetto; J. T., Conditioning *shampoo* composition, United States Patent 6,221,817; April 24, 2001, Assignee: The Procter & Gamble Company.
134. O'Lenick, Jr., A. J.; Smith, D. A.; Anderson, D,; United States Patent 6,881,710; April 19, 2005; assigned to Colonial Chemical Inc.
135. Wells, R.L.; Johnson, E.S.; U.S. Patent application 20030199403Al , October 23, 2003; Assignee: Procter & Gamble
136. Yeoh; T.; Coffindaffer; T. W.; Uchiyama; H.; Schroeder; J. G.; Okuyama; Y. '*Conditioning shampoo compositions having improved silicone deposition'* U.S. Patent 6,200,554; March 13, 2001.
137. Midha, S.; Hofrichter, B. D.; Thomson, S. R.; United States Patent Application 20030086896. May 8, 2003, assigned to The Procter & Gamble Company
138. Midha, S.; Hofrichter, B. D.; United States Patent Application 20030091521. May 15, 2003, assigned to The Procter & Gamble Company.
139. Wells, R.L.; Johnson, E.S.; Royce, D.A.; *US Pat Application 20030202952*, **Oct 30, 2003**.
140. Niemiec, H. Yeh; Gallagher, R.; Ho, K.L.; *US Pat Application 20030176303*, **Sep 18, 2003**.
141. Weimer D.R.; U.S. Patent 3,718,609. February 27, 1973; assigned to Continental Oil Company.
142. Olson, U.S. Patent 3,810,478; May 14, 1974; assigned to Colgate-Palmolive Company.
143. Williams, J. R.; Macaulay, E. W.; Aronson, M. P.; Massaro , M.; Salmon, T.; U.S. Patent. 6,429,177; August 6, 2002 ; assigned to Unilever Home & Personal Care USA, division of Conopco, Inc.
144. Littau, C.; Henning, T.; Porter, S.; Cooper, S.; U.S. Patent Application 20050215443; September 29, 2005. assigned to Clariant International Ltd.
145. Wei, K.S.; Thomas, C.P.; Taylor, R.A.; Tanner, P.R. ; Stella, Q.; Smith, E. D. III; United States Patent Application 20050100570; May 12, 2005, The Procter & Gamble Company.
146. Wagner, J. A.; Wei, K. S.; Smith, E. D. III; United States Patent Application 20050143269; June 30, 2005 The Procter & Gamble Company.
147. Wagner, J. A.; Wei, K. S.; Smith, E. D. III; United States Patent Application 20050192189; September 1, 2005; The Procter & Gamble Company.
148. Wagner, J. A.; Wei, K. S.; Smith, E. D. III; United States Patent Application 20050192188; September 1, 2005; The Procter & Gamble Company.

149. Midha, S.; Dunbar, J. C.; Hopkins, J. J.; Heinrich, J. M.; Comstock, B. G.; Wei, K. S.; United States Patent Application 20050143268; June 30, 2005; assigned to The Procter & Gamble Company.
150. Wagner, J. A.; Wei, K. S.; Smith, E. D. III; United States Patent Application 20050192187; September 1, 2005; The Procter & Gamble Company
151. Tsaur, L. S.; United States Patent 6,903,057; June 7, 2005; assigned to Unilever Home & Personal Care USA, division of Conopco, Inc.
152. Ugazio, S. P. J.; United States Patent Application 20050203215; September 15, 2005; Rohm & Haas Company.
153. Wells, R. L.; Johnson, E. S.; United States Patent 6,930,078; August 16, 2005; assigned to The Procter & Gamble Company
154. Qiu, Q.; Meyer, F.; Dave, R. M.; United States Patent Application 20050136026; June 23, 2005; Unilever Intellectual Property Group
155. Guth, J. J.; Spilatro, D. L.; Verdicchio, R. J.; United States Patent *4,435,300;* March 6, 1984; assigned to Johnson & Johnson Baby Products Company
156. Pratley, S. K.; United States Patent *6,001,787;* December 14, 1999; assigned to Helene Curtis, Inc.
157. Sivik, M. R.; Bodet, J. -F.; Kluesener, B. W.; Scheper, W. M.; Yeung, D. W.-K.; Bergeron, V.; United States Patent 6,900,172; May 31, 2005; assigned to Procter & Gamble Company.
158. Lange; Rainer; Rosato; Pietro; Wersuhn; Astrid; *Absorbent personal care and/or cleansing product for cosmetic and/or dermatological applications comprising at least one absorbent sheet*, United States Patent Application 20060019571, January 26, 2006
159. Hensen, Hermann; Lazarowitz, Virginia; Morris, Timothy; Kublik , Heike; Searle, Jackie; Leonard, Mark; *Impregnating solution for cosmetic cloths,* United States Patent Application 20060039956, February 23, 2006.
160. Erazo-Majewicz; Paquita; Modi; Jashawant M.; "*Functional systems with reduced odor cationic polygalactomannan",* United States Patent Application 20060046943, March 2, 2006.
161. Nakanishi, Tetsuo; Tanaka, Toshiki ; Tachibana, Kiyomi; United States Patent Application 20050008600, January 13, 2005.
162. Lu, Shaoxing; Kanji, Mohamed; United States Patent Application 20050020769, January 27, 2005: assigned to L'Oreal
163. Gonzalez; Anthony D; Hubbard; Barbara J.; Anderson; Glen; Kalafsky; Robert E.; *Topical skin care compositions,* United States Patent Application 20060045890; March 2, 2006.
164. O'Brien; Michael Joseph; *Process for making cosmetic compositions using polyether siloxane copolymer network compositions;* United States Patent Application 20060052521; March 9, 2006.

165. O'Brien, Michael Joseph; *Polyether siloxane copolymer network compositions;* United States Patent Application 20060052520; March 9, 2006.
166. Bosman, Anton Willem; Janssen, Hendricus Marie; Van Gemert, Gaby Maria Leonarda; Versteegen, Ronny Mathieu; Meijer, Egbert Willem; Sijbesma, Rintje Pieter; *Siloxane polymers with quadruple hydrogen bonding units;* United States Patent Application 20060018856; January 26, 2006
167. Castrogiovanni, Anthony; Barone, Salvatore J.; Krog, Ann; McCulley, Marion L.; Callelo, Joseph F.; United States Patent 5,505,937; April 9, 1996; assigned to Revlon.
168. Nishio, Masaya; *Liquid skin protective composition*; United States Patent Application 20060039885, February 23, 2006
169. Powell, Virginia V.; Kasson, Amy-Elizabeth; Kilgour, John A.; Kuo, An-Li; *Non-aqueous silicone emulsions*; United States Patent 6,060,546; May 9, 2000; assigned to General Electric.
170. Feng, Qian Jane; Lin, Zuchen; Hill, Randal Myron; *Clear silicone microemulsions formed spontaneously;* United States Patent 6,998,424, February 14, 2006; assigned to Dow Corning
171. Nakanishi, Tetsuo; Sakuta, Koji; Tachibana, Kiyomi; *Novel organopolysiloxane polymer, pasty composition, and cosmetic preparation containing the composition*, United States Patent Application 20060034875, February 16, 2006.
172. Fabry, Bernd; Koester, Josef; Hensen, Hermann; Schmid, Karl Heinz; *Cosmetic preparations containing silicone compounds and esters of hydroxycarboxylic acids and alk(en)yl oligoglycosides*; United States Patent 7,001,591, February 21, 2006, assigned to Cognis
173. Pickering, T.; *J Chem. Soc* . 91 (1907).
174. Bancroft, W.; Theory of emulsification III, *J Phys Chem,* 16, 475 (1912)
175. Corcorran, S; Lochhead, R; McKay, T; Particle-Stabilized Emulsions: A Brief Overview; *Cosmetics & Toiletries,* September, 2004
176. Gers-Barlag, Heinrich; Muller, Anja; United States Patent 6,838,088; January 4, 2005. assigned to Beiersdorf AG.
177. Lennon, Paula; Lorant, Raluca; United States Patent 6,855,311, February 15, 2005; assigned to L'Oreal.
178. Allgaier, J.; Poppe, A.; Willner, L.; Richter, D.; Synthesis and Characterization of Poly (1, 4 .isoprene–*b*–(ethylene oxide) and poly [(ethylene–*co*–propylene-*b*–ethylene oxide)] Block Copolymers, *Macromolecules*, (1997), Vol. 30, 1582.
179. Laufer *et al.*, *Cosmetics and Toiletries*, 111, (Jun. 1996), 92.
180. Simonnet, Jean-Thierry; Biatry, Bruno; Saint-Leger, Didier; United States Patent Application 20050031699, February 10, 2005; assigned to L'Oreal.

181. Brunauer, Emmet and Teller; J. Amer. Chem. Soc. **1938**, *60*, 309.
182. Verloo, Aurore; Fontaine, Jacqueline; United States Patent Application 20040265346; December 30, 2004
183. Do Amaral Martins; Marcelo; Asua; Jose M; De Brouwer; Hans; Van Es; Steven; *Aqueous polymer dispersion and process;* United States Patent Application 20060052529; March 9, 2006. assigned to Surface Specialties S.A.
184. L'Alloret, Florence; United States Patent Application 20050008605, January 13, 2005, assigned to L'Oreal.
185. Wanka, G.; Hoffman, H.; Ulbricht, W.; *Colloid Polym. Science*, 1990, *268*, 101.
186. Linse, P.; Malmsten, M.; *Macromolecules,* **1992**, *25*, 5434.,
187. Mercier, Jean-Michel; United States Patent Application 20050019352, January 27, 2005, filed by Rhodia Inc.
188. Ron; Eyal S.; Bromberg; Lev; Temchenko; Marina; *End modified thermal responsive hydrogels;* United States Patent 6,316,011, November 13, 2001; assigned to Madash LLC
189. Olivieri; Laetitia; Ron; Eyal S.; *Thermally reversible water in oil in water emulsions;* United States Patent 6,995,209; February 7, 2006, assigned to Madash LLC
190. Bitler; Steven P; Taft; David D; *Polymeric thickeners for oil-containing compositions;* United States Patent 6,989,417; January 24, 2006; assigned to Landec Corporation
191. Ahmad, Nawaz; Lamb, Christopher Scott; Lonardo, Emilia Casilio; Joyce, Michael E.; Patel, Kalpana J.; Heda, Pavan Kumar; United States Patent Application 20050042248, February 24, 2005, filed by Johnson & Johnson.
192. Ahmad, Nawaz; Lamb, Christopher Scott; Lonardo, Emilia Casilio; Joyce, Michael E.; Patel, Kalpana J.; Heda, Pavan Kumar; United States Patent Application 20050042249, February 24, 2005, filed by Johnson & Johnson.
193. Fossel; Eric T.; *Topical delivery of L-arginine to cause tissue warming;* United States Patent 5,895,658, April 20, 1999
194. Menkart et al, U.S. Pat. No. 3,250,680
195. Hertzenberg; Elliot P., *Zeolite for personal care products*; United States Patent 4,626,550; December 2, 1986; assigned to PQ Corporation
196. Gott; Robert Edward ; Slavtcheff; Craig Stephen ; Znaiden; Alexander Paul; Crotty; Brian Andrew; *Cosmetic compositions with self-warming component.* United Start's Patent 6,287,580; September 11, 2001; assigned to Unilever
197. Church, John A.; Mehta, Umesh J.; *Autogenously heated liquid soap composition;* United States Patent 4,839,081; June 13, 1989; assigned to Colgate-Palmolive

198. Barnhurst, James D.; Shumway, Durland K; *Method of treating skin and hair with a self-heated cosmetic;* United States Patent 4,110,426; August 29, 1978; assigned to Colgate-Palmolive
199. Ahmad, Nawaz; Lamb, Christopher Scott; Lonardo, Emilia Casilio; Joyce, Michael E.; Patel, Kalpana J.; Heda, Pavan Kumar; *Warming and nonirritating lubricant compositions and method of comparing irritation,* United States Patent 7,005,408, February 28, 2006; assigned to McNeil-PPC.
200. Shefer, Adi; Shafer, Samuel David; *Self-warming or self-heating cosmetic and dermatological compositions and method of use;* United States Patent Application 20060039886; February 23, 2006.
201. Ferrari, Veronique; Tournilhac, Florence; *Compositions Comprising At Least One Organic Polymer And A Mixture of Volatile And Nonvolatile Solvents;* United States Patent Application 20060039881; February 23, 2006, assigned to L'Oreal.
202. White, John H.; *Alcohol-based antimicrobial compositions;* United States Patent 5,288,486; February 22, 1994: assigned to Calgon Corporation.
203. Jampani, Hanuman B. (Grapevine, TX); Newman, Jerry L. (Arlington, TX); Newman, Anthony W.; *Alcohol based anti-microbial compositions with cosmetic appearance;* United States Patent 5,997,893: December 7, 1999; assigned to Ethicon.
204. Greten, Zachariah C.; Kaiser, Nancy-Hope E.; Klein, Daniel A.; *Enhanced activity alcohol-based antimicrobial compositions;* United States Patent Application 20060039942; February 23, 2006.
205. Hart, Scott A.; Zeh, Karin; Machleidt, Thomas; *Compositions having antimicrobial activity and uses thereof;* United States Patent Application 20050282755, December 22, 2005; assigned to Ansata Therapeutics
206. Rothbard, Jonathan B; Wender, Paul A.; McGrane, P. Leo; Sista, Lalitha V. S.; Kirschberg, Thorsten A. *Compositions and methods for enhancing drug delivery across and into epithelial tissues* United States Patent *6,593,292,* July 15, 2003; assigned to CellGate.
207. Scholz, Matthew T.; *Cationic antiseptic compositions and methods of use,* United States Patent Application 20060051385, March 9, 2006; assigned to 3M Innovative Properties Company.
208. Kawasaki, Yuji; Hori, Michimasa; Yamamoto, Yuichi; Hiraki, Jun; *Cosmetic composition and production thereof;* United States Patent Application 20060018867; January 26, 2006; assigned to Ichimaru Pharcos Co.
209. Hessefort, Yin Z.; Morris, John D.; Carlson, Wayne M.; Wei, Mingli; Quadir, Murat; Brammer Jr., Larry E.; *Water soluble monomers and polymers for protecting substrates from ultraviolet light;* United States Patent 7,008,618; March 7, 2006; assigned to Nalco.

Polymer Architecture and Film Performance

Chapter 2

Effect of Hydrophobic Substitution on Cationic Conditioning Polymers

Susan L. Jordan[1], R. Lowell Kreeger[2], Xiaodong Zhang[2], Tatiana V. Drovetskaya[2], Calvin B. Davis[2], Jennifer L. Amos[2], Stephanie E. Gabelnick[3], Shuiqin Zhou[4], Wingkin Li[2], Edward F. DiAntonio[2], and Anita A. Protonentis[2]

[1]The Dow Chemical Company, 171 River Road, Bound Brook, NJ 08854
[2]Amerchol Corporation, 171 River Road, Bound Brook, NJ 08854
[3]Nalco Company, 1601 W. Diehl Road, Naperville, IL 60563
[4]Department of Chemistry, College of Staten Island, and The Graduate Center, The City University of New York, 2800 Victory Boulevard, Staten Island, NY 10314

Cationic cellulosic polymers are formulated with anionic surfactants in shampoos and body washes as conditioners and deposition aids. During dilution, the polymer and surfactant undergo a series of association and microstructure changes: from a condensed network to globular aggregation to inter-chain association with micelles. Cationic polymer structure impacts the polymer/surfactant interaction. The addition of hydrophobic chains on the polymers significantly changes deposition and conditioning properties. In general, deposition of benefit agents from shampoos and body washes increases in the presence of these modified polymers. Conditioning in clear formulations also improves. These performance changes depend on the absolute molar substitution of the hydrophobic moiety as well as the surfactant system. The presence of even very low levels of hydrophobe alters interfacial reactivity with surfactants and surfaces influencing both coacervate formation and deposition.

Polyquaternium–10 (PQ-10) is the INCI name for cationic (trimethyl ammonium) hydroxyl ethyl cellulose (HEC) polymers that are used in shampoo systems to impart conditioning properties to hair including good feel properties and manageability. More recently, these polymers are used in body wash formulations to improve skin feel during washing and after drying and improved deposition of oils. Nomenclature for commercial PQ10 polymers reflects the relative molecular weight and charge density. The first two letters represent charge density (JR:~1.7% nitrogen, LR: ~1.0% Nitrogen). The following numbers/letters indicate relative molecular weight (400:medium, 30M-:high).

In shampoo and body wash formulations, the cationic polymer and surfactant exist together in a single phase. Upon dilution during washing and rinsing, cationic HEC polymers form a complex with the surfactant system that phase separates from the bulk solution. This phase separation (or coacervation) is a well-known phenomenon known as the "Lochhead Effect" (1-3). The coacervate is a gel like phase that contains a high level of cationic polymer that deposits on negatively charged substrates, forming a clear film. The haze formed by coacervate formation is readily measured. In addition, coacervates aid in the deposition of insoluble actives, such as silicone (4,5). Deposition efficiency is dependent on polymer molecular weight as well as cationic charge density. Recent work by Zhou, et al. (6) investigates molecular weight effects and ethylene oxide chain length on the phase behavior of cationic hydroxyethyl cellulose at different concentrations of sodium dodecyl sulfate (SDS). These phase changes include a condensed network, globular aggregation (coacervation) and inter-chain association with micelles (Figure 1).

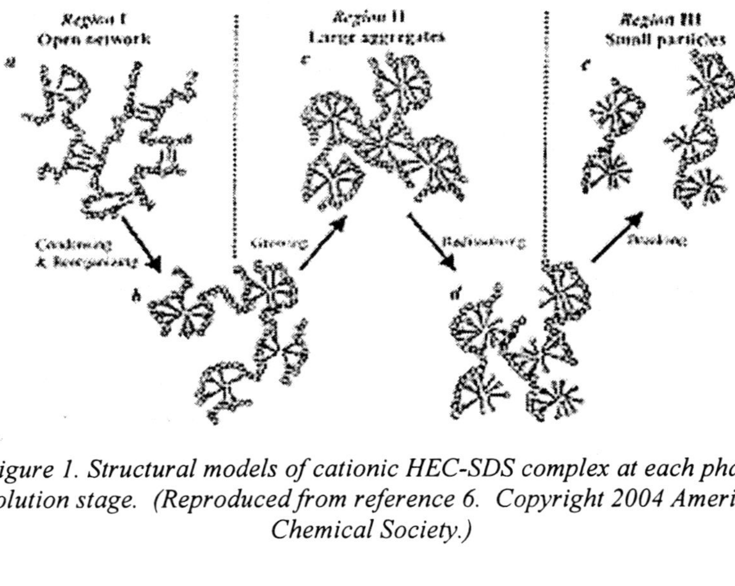

Figure 1. Structural models of cationic HEC-SDS complex at each phase evolution stage. (Reproduced from reference 6. Copyright 2004 American Chemical Society.)

Zhou, et al. show at low surfactant, the polymer is soluble with a hydrodynamic radius (Rh) about 100 nm near the phase boundary. As surfactant concentration increases, Rh increases and phase separation occurs. As the upper boundary is approached, particle size is maximized. Addition of surfactant to concentrations above the upper phase boundary indicates resolubilization of the polymer and decreased Rh. The complex microstructure relationships between polymer and surfactants predict additional functional and performance effects upon the addition of hydrophobic chains onto the polymer backbone (Figure 2).

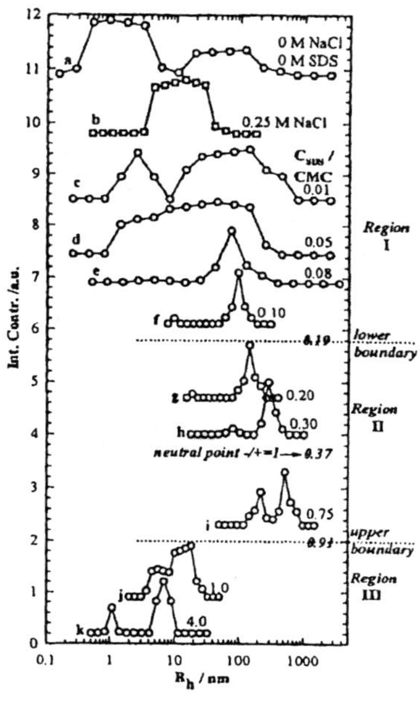

Figure 2. Apparent hydrodynamic radius (Rh) distribution of 0.2 wt % EO22MW110 solution in0.25 M NaCl (b) and in differently concentrated SDS solution (c-k). (Reproduced from reference 6. Copyright 2004 American Chemical Society.)

Materials and Methods

All materials were used as received:

Cationic polymers: PQ10 and PQ67 from Amerchol Corporation

Surfactants (Cognis): Standapol ES-2 (sodium laureth sulfate, SLES-2),

Velvetex BK-35 (cocoamidopropyl betaine, CAPB), Velvetex CDC (disodium cocamphodiacetate, DSCADA)

Silicone: Dow Corning 2-1664 emulsion

Hair: International Hair Importers and Products

Vitro-Skin®: IMS Inc.

Wet combing: Wet combing is measured using a Dia-Stron Miniture Tensile Tester. The force required to pull a comb through a wet hair tress treated with surfactant alone and then surfactant plus polymer is measured. Five grams of hair is washed with 0.5 grams of a shampoo, and rinsed at constant temperature. The difference in force required to comb the hair tress is wet comb reduction.

Coacervate study: Haze measurements were performed using a nephalometer or a Nippon Denshoku 300A hazemeter 2 minutes after diluting formulation 2.5 to 20 times with water and mixing at room temperature.

Silicone deposition: The amount of silicone deposited on hair after shampooing was measured. Five grams of hair was washed with 0.5 grams of a shampoo, and rinsed at constant temperature. Hair was extracted with a 1:1 mixture of methyl butyl ketone and toluene. Atomic absorption spectroscopy was used to measure silicone content; reporting μg silicone/g hair.

Sunflower seed oil deposition: A 3-6 cm piece of Vitro-skin® was washed with 0.15 mL body wash and rinsed at constant temperature. Dried skin was extracted with heptane and analyzed for SSO using gas chromatography.

Results and Discussion

The effect of molecular weight and charge density of cationic HEC polymers (Polyquaternium 10 or PQ10) on silicone (0.5 μ dimethicone) deposition from a simple shampoo formulation (15.5% SLES-2 / 2.6% DSCADA, 0.25% polymer, 1% silicone) is shown in Figure 3. Both parameters affect the overall performance of the polymers in depositing silicone. This data is consistent with coacervate formation as a function of dilution level.

The haziness of the diluted solution was monitored as an indicator of coacervate formation and was measured with a Nephalometer 2 minutes after dilution. In this case, the coacervate peak-height increases with molecular weight, suggesting more coacervate leads to increased deposition and may simply be a function of the increased ability of the larger polymer to flocculate the insoluble silicone onto the hair rather than a change in interactions in solution. Coacervate peaks shift significantly (from 5 to 10 fold dilution), however, as the charge density decreases (Figure 4). This shift is independent of surfactant base (Shampoo 1: 15.5% sodium laureth sulfate (SLES-2) / 2.6% cocoamidopropyl betaine (CAPB), Shampoo 2: 15.5% SLES-2 / 2.6% disodium cocoamphodiacetate (DSCADA) Data not shown). Differences in silicone deposition coupled with coacervate peak shifts at varying charge density suggest

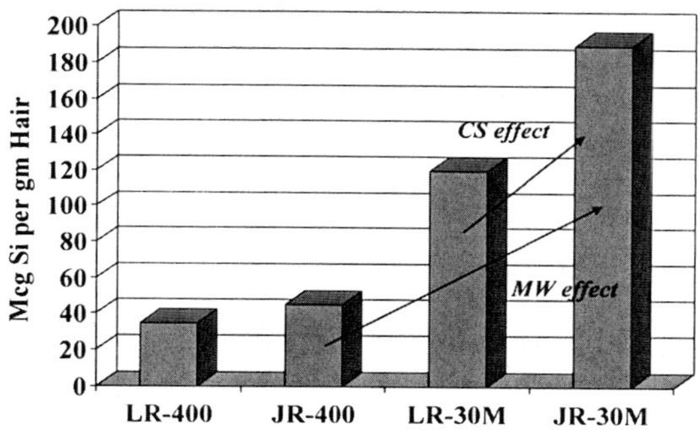

Figure 3. Silicone deposition onto European brown hair using PQ10. Low molecular weight polymers:- 400, high molecular weight polymers: -30M; low charge density: LR, high charge density: JR. (15.5% SLES-2 / 2.6% DSCADA, 0.25% polymer, 1% silicone)

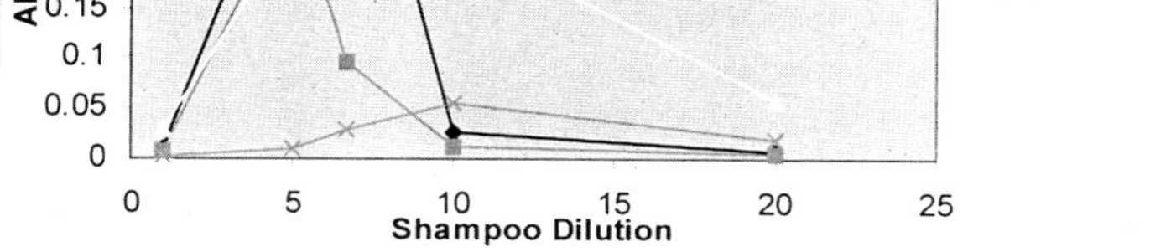

Figure 4. Coacervate haze curves of PQ10 polymers measured using a Nephalometer. Low molecular weight polymers:- 400, high molecular weight polymers: -30M; low charge density: LR, high charge density: JR. (15.5% SLES-2 / 2.6%CAPB, 0.5% polymer) (See page 1 of color inserts.)

the interaction of the polymer with surfactant is highly dependent on charge density.

Wet combing studies are used to measure relative conditioning of the hair. The force required to pull a comb through wet hair is measured with a Diastron and compared to a control. The more wet comb force reduction, the better the conditioning. Polymer dependence on conditioning effects, as measured by wet combing, show that as molecular weight increases, wet combing increases. In contrast, while charge increases, wet combing decreases. The effect of molecular weight, however, appears to partially compensate for the effect of charge on overall wet combing (Figure 5).

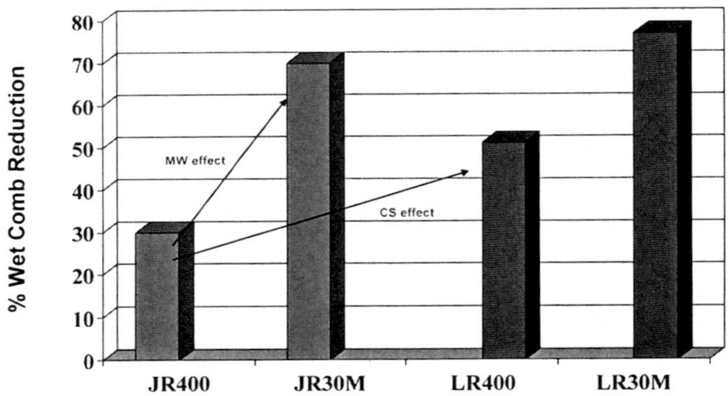

Figure 5. Wet combing reduction of single bleached hair washed with shampoo containing PQ10 polymers. Low molecular weight polymers:- 400, high molecular weight polymers: -30M; low charge density: LR, high charge density: JR. (15.5% SLES-2 / 2.6% DSCADA, 0. 5% polymer)

Based on the data in Figures 3-5, both silicone deposition and conditioning are favored by higher molecular weight polymers, regardless of charge substitution. Interestingly, early peak dilution (higher charge) favors silicone deposition while late peak dilution and wide coacervation zone (lower charge) favor conditioning. PQ10 also effects sunflower seed oil (SSO) deposition from body wash formulations. Similar to shampoo systems, increased molecular weight and charge density result in increased SSO deposition onto skin (data not shown).

According to Zhou, et al (6) differences in polymer surfactant microstructures on dilution are apparent with cationic HEC polymers containing lower levels of ethylene oxide (EO) substitution from typical PQ10. Less EO suggests the polymer will be less hydrophilic and have more hydrophobic character. The change in microstructure observed with low EO polymers suggests increased hydrophobicity of cationic HEC will affect interaction with surfactants and hence polymer performance. Zhou, et al. determined polymers with low EO substitution have a broadened coacervate zone (suggesting better wet combing) and increased Rh (suggesting better deposition). As was shown in Figure 2, the largest Rh occurs at high surfactant concentration (low dilution). Since lower peak dilution corresponds to better deposition (Figures 3 & 4), it is possible a more hydrophobic polymer can both increase peak area (maximize wet comb) and maintain maximum peak dilution at low dilution levels (maximize deposition).

In order to understand the effect of hydrophobicity on cationic HEC, a new polymer family, Polyquaternium 67 (PQ 67), was developed. PQ 67 is a high viscosity quaternized hydroxyethyl cellulose (HEC) polymer with cationic substitution of trimethyl ammonium and dimethyldodecyl ammonium (Figure 6). For experiments in shampoo systems, cationic substitution was fixed at ~ 0.2, which corresponds to a weight-percent nitrogen of ~1.0%, while substitution of hydrophobic dimethyldodecyl ammonium is held at hydrophobic substitution < 0.01. Four hydrophobic levels were tested with designations SL-5, SL-30, SL-60, and SL-100, where the molar substitution of relative hydrophobe increases with increasing number from 5 to 100. The results indicate the addition of even low concentrations of hydrophobe leads to significant differences in performance.

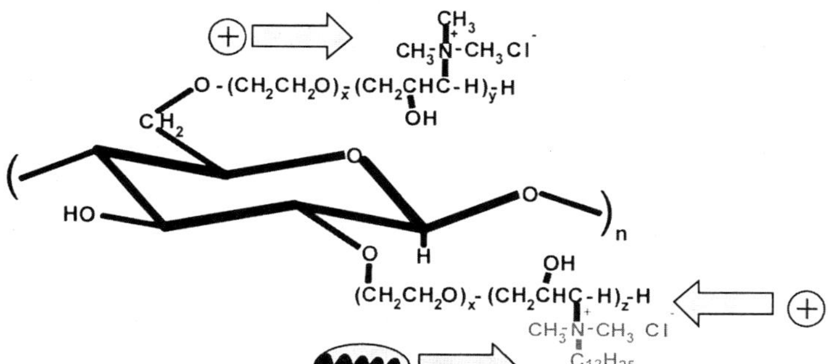

Figure 6. Polyquaternium 67: cationic HEC containing Hydrophobic substitution < 0.01 dimethyldodecyl ammonium.

A significant increase in wet combing performance is seen in shampoos formulated with hydrophobically modified polymers. This performance

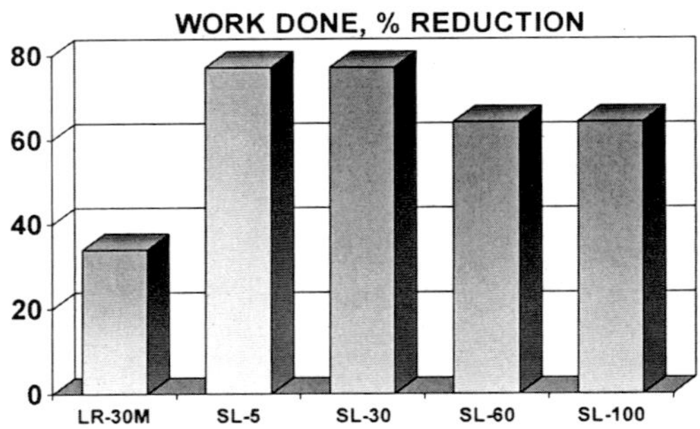

Figure 7. Wet combing force reduction of single bleached European brown hair using hydrophobically modified cationic HEC polymers. 15.2% SLES-2 /2.56% DSCADA /0.3% polymer.

enhancement does not continue to increase with higher levels of hydrophobic substitution (SL-60 and SL-100) suggesting performance differences are not due to molecular weight effects, but to the hydrophobe itself. (Figure 7)

Silicone deposition onto hair is shown in Figure 8. Similar to wet combing data, silicone deposition increases in the presence on hydrophobe. Given the low level of hydrophobe, there is not a significant increase in molecular weight of these polymers. The near doubling of silicone deposition cannot be explained by simple increased flocculation and bridging caused by a larger polymer. It is more likely the hydrophobe is changing the interaction of the polymer with the surfactant and polymer with silicone.

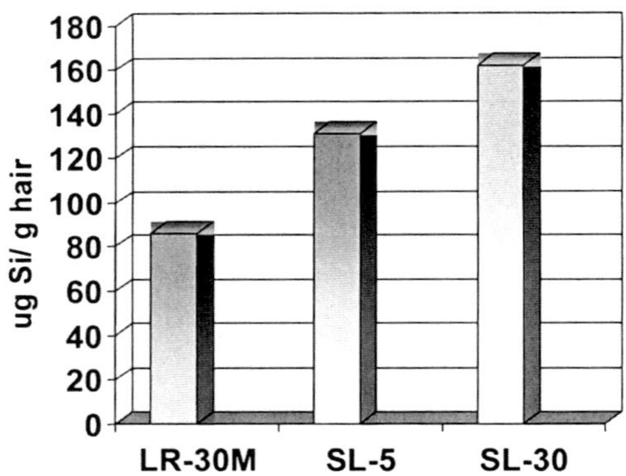

Figure 8. Silicone deposition on European brown hair of hydrophobically modified cationic HEC polymers. 15.2% SLES-2 / 2.56% DSCADA/0/3% polymer/0.25% polymer/1% dimethicone blend.

PQ67 polymers were also investigated in body wash systems (11% SLES-2, 4% CAPB, 15% sunflower seed oil, and 0.5% polymer). These polymers use the designation SK. The SK polymers tested vary charge substiution rather than hydrophobic density level (Figure 9). The effect of molecular weight on deposition of sunflower seed oil onto *Vitro*-skin is shown in Figure 10 using commercial PQ10 polymers. As with silicone deposition on hair, higher molecular weight polymer of either low or high charge density (LR and JR polymers, respectively) exhibits higher sunflower seed oil deposition. This increase in deposition may be related to large polymer bridging effects. Figure 10 also shows the effect of charge substitution on deposition. The higher charged JR30M (~1.7% N) deposits more oil than the lower charged LR30M (~1.0% N).

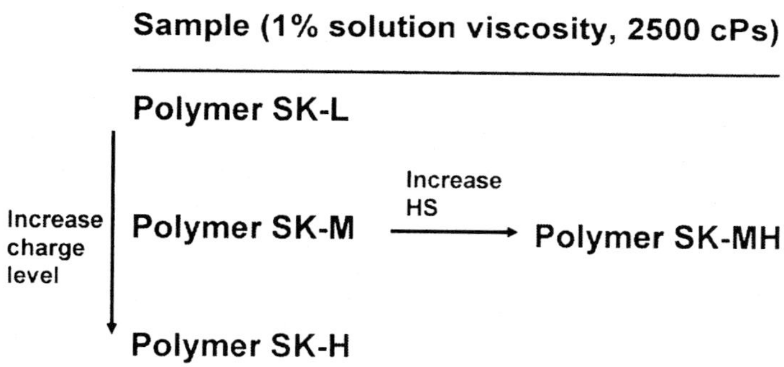

Figure 9. SK polymers used in body wash formulations. Hydrophobe density for SK-L, SK-M, & SK-H similar to SL-5; SK-MH similar to SL-30.

Additional charge substitution effects for SSO deposition are shown in Figure 11. In the presence of hydrophobe, rather than simply increasing with increasing charge, deposition actually peaks between LR30M (1.0% N) and JR30M (1.7% N).

As predicted from the low EO substitution studies and the results above, the presence of hydrophobe shifts coacervate curves significantly (Figure 12). The peak coacervate formed with SK-M (1.25% nitrogen) corresponds to peak coacervate formed with JR30M (1.75% nitrogen). The addition of hydrophobe shifted the peak coacervate to early dilution, leading to increased deposition. A similar peak shift is seen with SK-L (1.05% nitrogen) compared with essentially the same nitrogen level of LR-30M (0.95% nitrogen). The only difference being the presence of hyrdophobe on SK-L. Both SK polymers show significantly broader coacervate curves, therefore accounting for the improved wet combing as well. This non linear effect of charge and hydrophobe substitution suggests

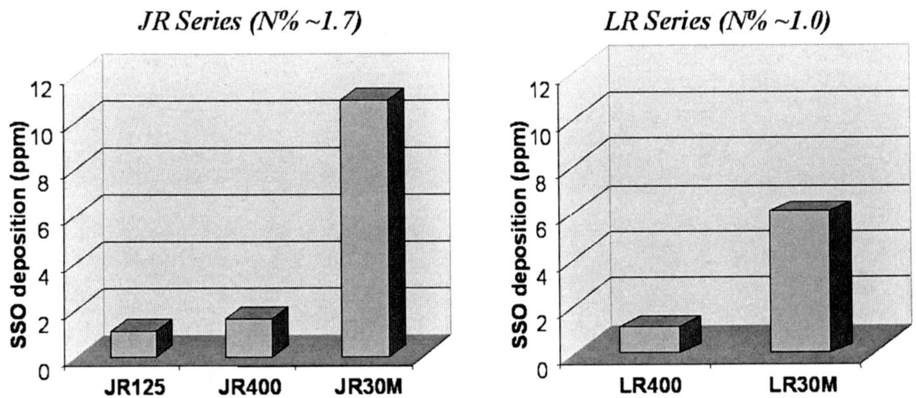

Figure 10. Molecular weight effects of sunflower seed oil deposition on Vitro skin: 11% SLES-2, 4% CAPB, 15% sunflower seed oil, 0.5% polymer. (MW: 125<400<30M. Charge: LR~1.0%, JR ~1.7%)

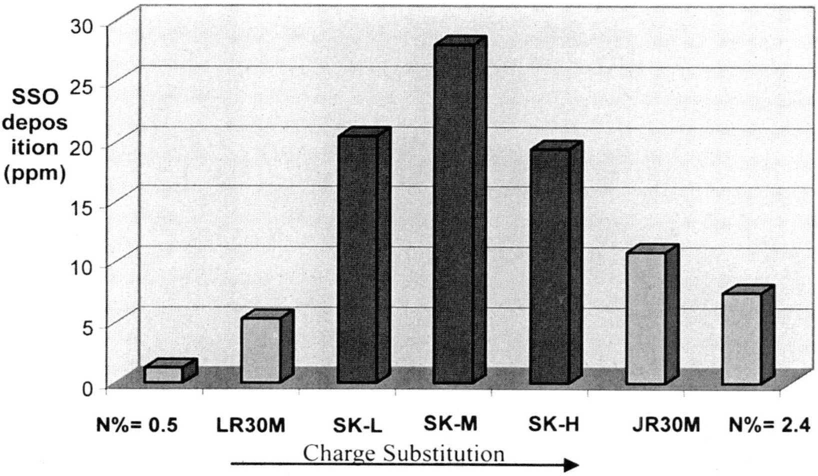

Figure 11. Effect of charge density in the presence of hydrophobe on sunflower seed oil deposition on in Vitro skin. 11% SLES-2, 4% CAPB, 15% sunflower seed oil, 0.5% polymer.(SK-L, 1.05%N; SK-M, 1.25%N; SK-H, 1.45% N)

specific structure/function relationships on the interaction of the polymer and surfactant which influence overall formulation performance.

Conclusion

The cationic HEC polymer architecture was found to have a significant effect on the formation and dissolution of surfactant-polymer coacervates, which were shown to be critical in deposition of water insoluble actives from shampoos and body wash formulations. A new compositional variant of cationc HEC polymers was synthesized by adding low levels of hydrophobic substitution (dodecyl chain). These hydrophobic chains had a pronounced affect on the deposition of insoluble actives presumably by associating with the surfactants through hydrophobic interaction, altering the interactions, and affecting overall formulation performance. The coacervate curves shifted significantly to the left (lower dilution) while continuing over a broad zone. With the addition of low levels of hydrophobic substitution, significant improvement in coacervate formation, and silicone and sunflower seed oil deposition were achieved. The data also reveal that coacervate curves can be used to predict overall polymer performance in rinse off systems.

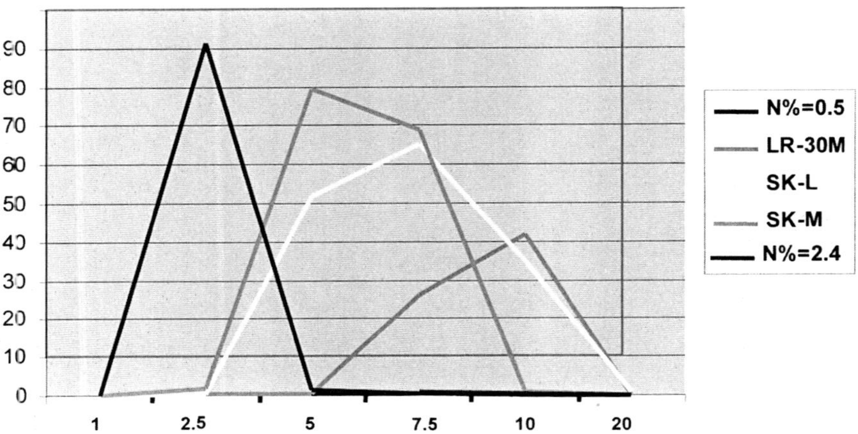

Figure 12. Coacervate haze curves at different charge density measured using a haze meter. 11% SLES-2, 4% CAPB, 0.5% polymer. (See page 1 of color inserts.)

References

1. Lochhead, R.Y., Shampoos, in *The Chemistry and Manufacture of Cosmetics*, 3rd ed., M.L. Schlossman, Ed., Allured Publishing Corporation, Carol Stream IL, 2002, vol 2 – formulating, pp 277–326.
2. Lochhead, R.Y., *Soap, Cosmetics, and Chemical Specialties,* **Oct. 1992**, pp 42–49.
3. Goddard, E.D., Polymer/Surfactant Interaction in Applied Systems, in *Principles of Polymer Science and Technology in Cosmetics and Persoanl Care*, E.D., Goddard and J.V. Grubber, Eds., Marcel Dekker, New York, 1999, p. 181.
4. Gamez-Garcia, M., *IFSCC*, **2001**, vol. 4, pp. 99–107.
5. Li, W., & Jordan, S.L.P., *Cosmetics and Toiletries Manufacture Worldwide*, **2003**, pp. 1–4.
6. Zhou, S., Xu, C., Wang, J., Golas, P., Batteas, J., & Kreeger, L., *Langmuir*, **2004**, vol. 20, pp. 8482–8489.

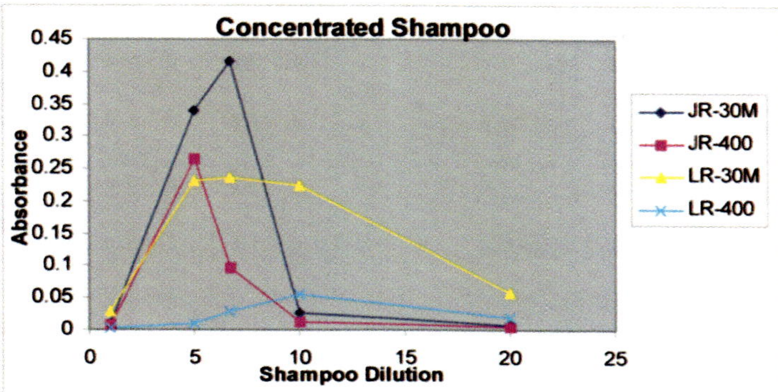

Figure 2.4. Coacervate haze curves of PQ10 polymers measured using a Nephalometer. Low molecular weight polymers: - 400, high molecular weight polymers: -30M; low charge density: LR, high charge density: JR. (15.5% SLES-2 / 2.6%CAPB, 0.5% polymer)

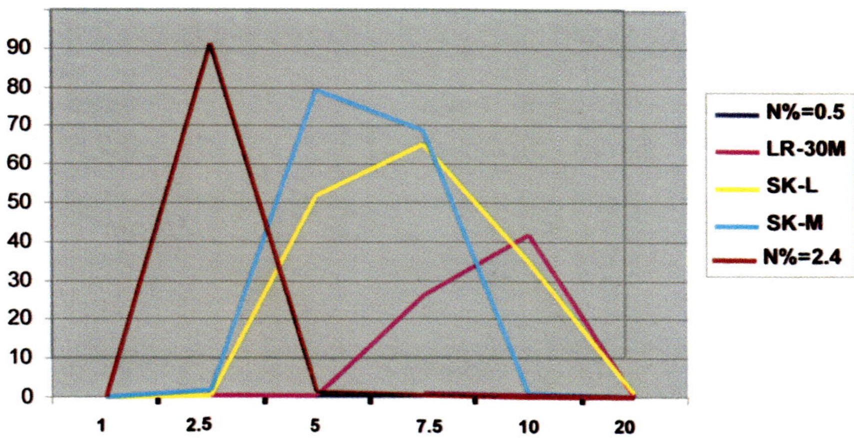

Figure 2.12. Coacervate haze curves at different charge density measured using a haze meter. 11% SLES-2, 4% CAPB, 0.5% polymer.

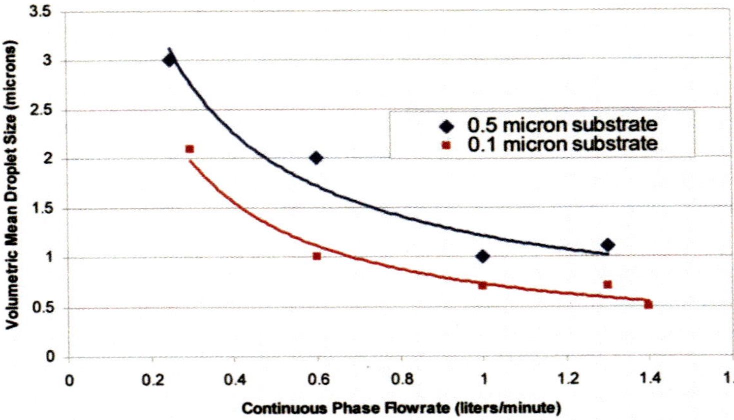

Figure 4.7. Emulsion mean droplet size as a function to continuous phase flowrate and a function of the substrate pore size for a constant oil flowrate of 10 mL/min (Reproduced with permission from Cosmetics and Toiletries *Copyright 2005.)*

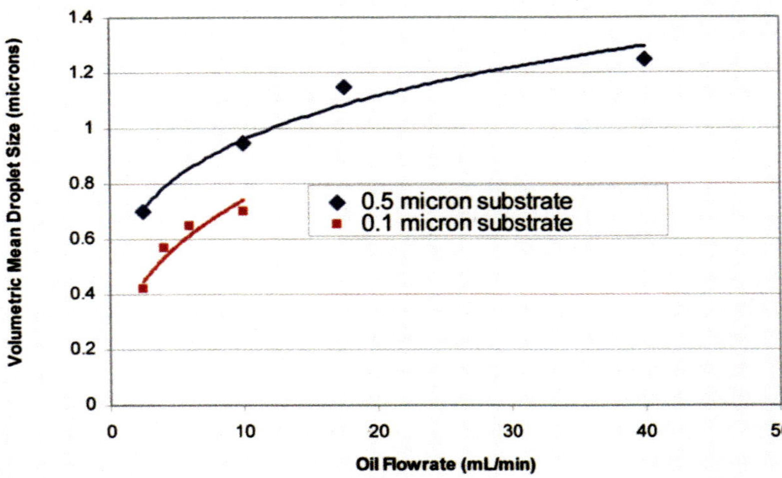

Figure 4.8. Emulsion mean droplet size as a function of the oil phase flowrate and a function of the substrate pore size for a constant water flowrate of 1 LPM (Reproduced with permission from Cosmetics and Toiletries *Copyright 2005.)*

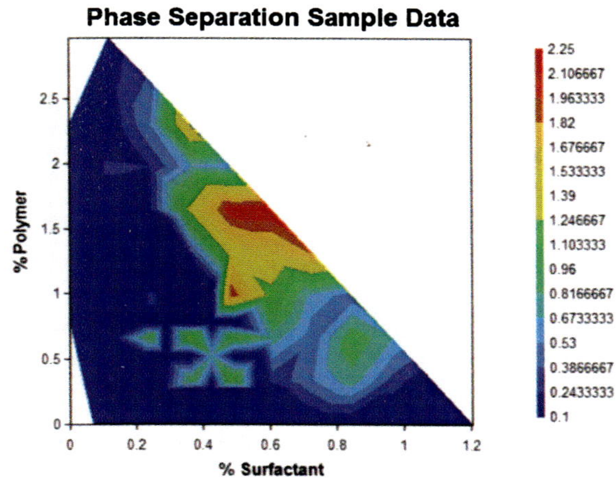

Figure 5.1. Sample contour phase diagram.

Method for Vinyl Pyridine Polymers

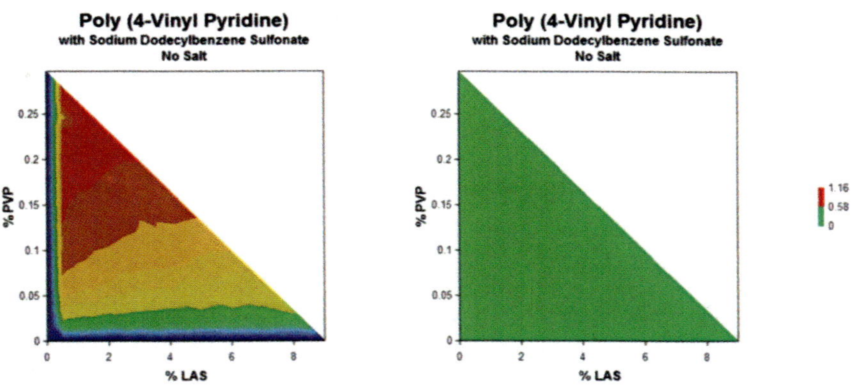

No salt

Figure 5.2. Reproducibility of high-throughput screening methods for synthetic polymer-surfactant interaction contour phase diagrams.
Continued on next page.

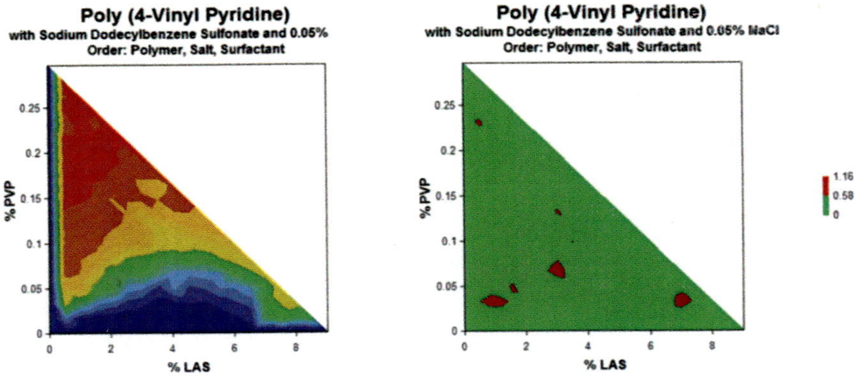

Polymer, Electrolyte, Surfactant

Figure 5.2. Continued.

Method for Cellulosic Polymers

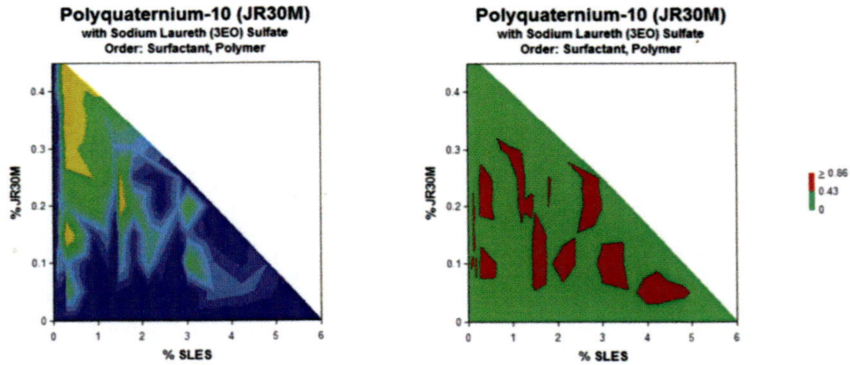

No salt

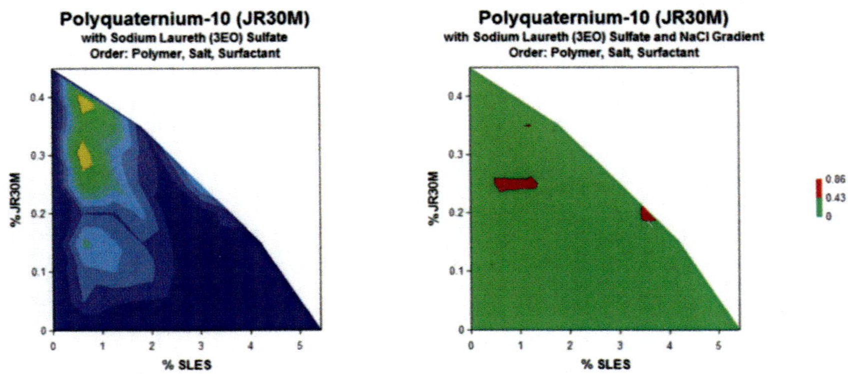

Polymer, Electrolyte, Surfactant

Figure 5.3. Reproducibility of high-throughput screening methods for cellulosic polymer-surfactant interaction contour phase diagrams.

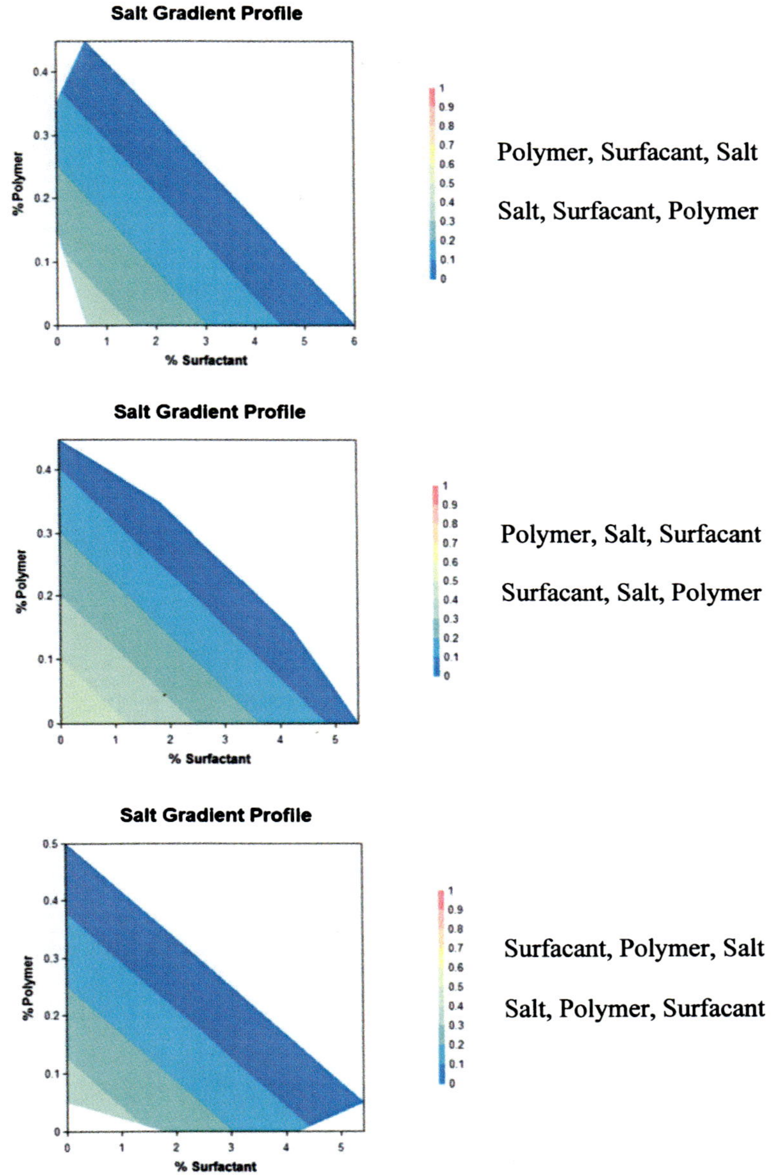

Figure 5.4. Salt gradient profiles for cellulosic polymer-SLES-NaCl investigations. The addition orders associated with each salt profile are listed to the right of the salt gradient profile phase diagrams.

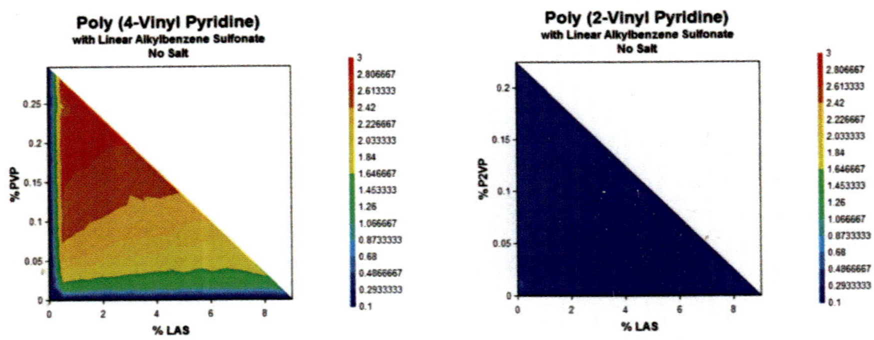

(a) (b)

Figure 5.6. Contour phase diagrams of (a) P4VP and (b) P2VP with LAS.

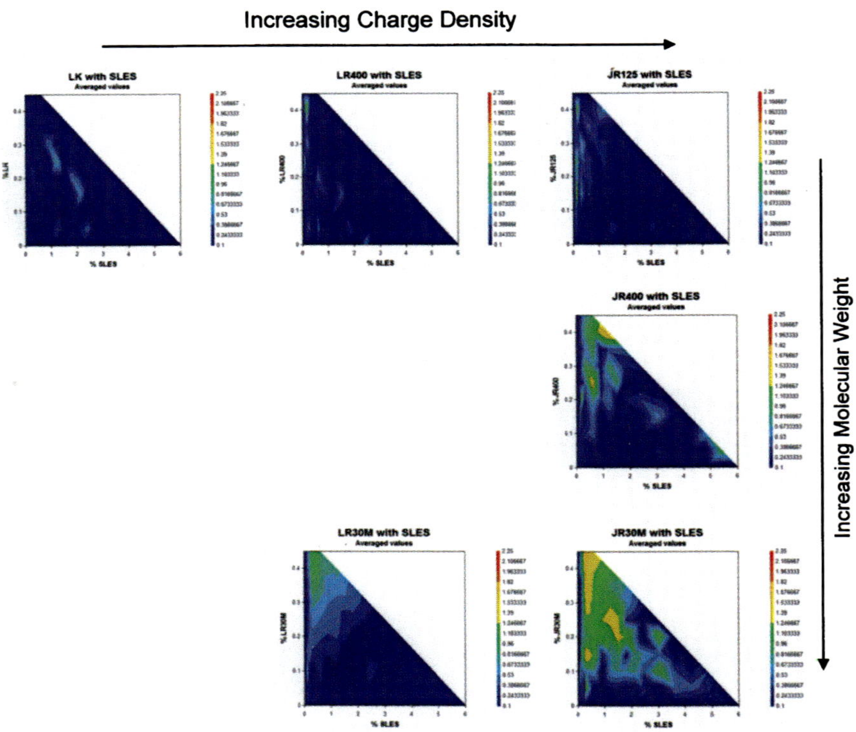

Figure 5.7. Contour phase diagrams of polyquaternium-10 polymers with SLES.

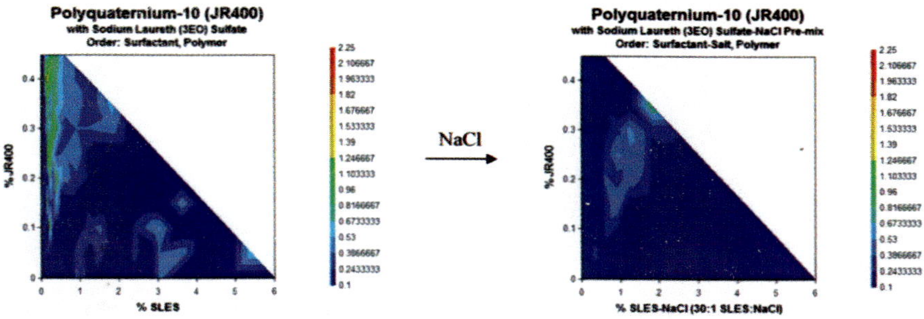

Figure 5.8. Effect of NaCl on phase separation (NaCl premixed with SLES).

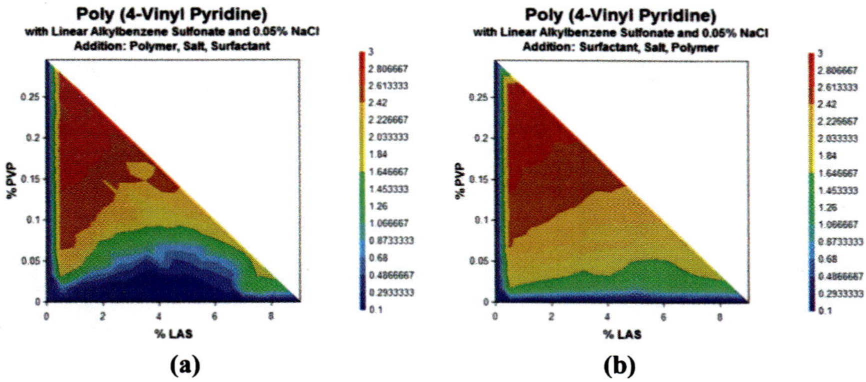

Figure 5.9. Phase diagrams for P4VP with LAS (a) Polymer, Salt, Surfactant and (b) Surfactant, Salt, Polymer.

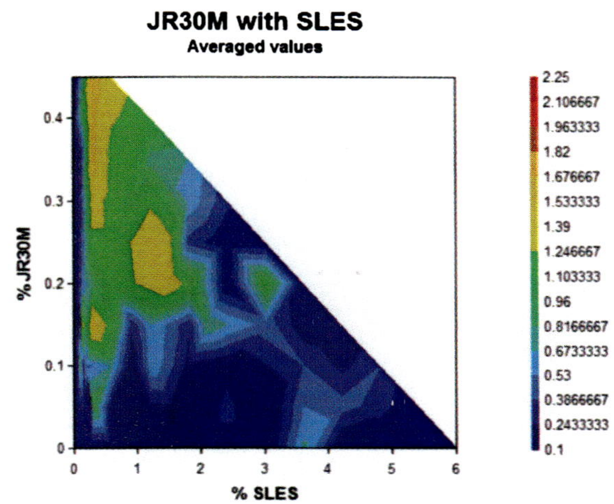

Figure 5.10. Contour phase diagram of JR30M-SLES with no salt.

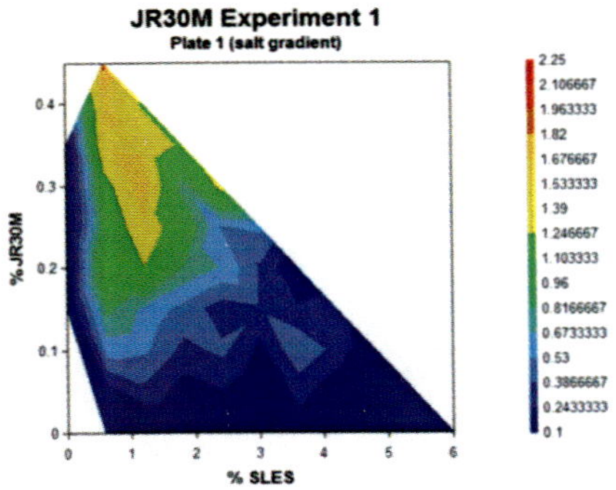

(a)

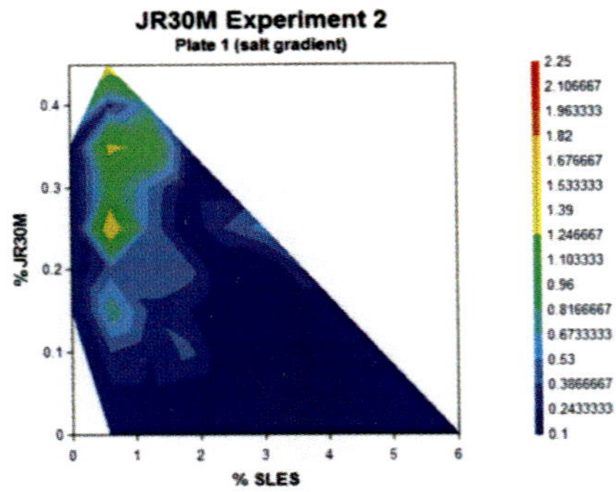

(b)

Figure 5.11. Contour phase diagrams for JR30M-SLES-NaCl with (a) Salt, Surfactant, Polymer and (b) Polymer, Surfactant, Salt addition orders.

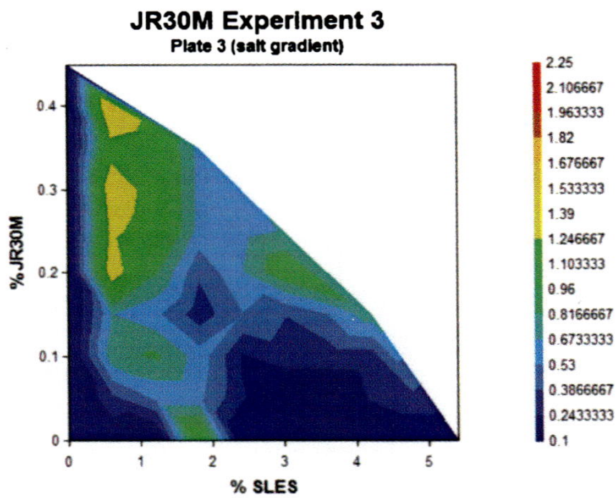

Figure 5.12. Contour phase diagram for JR30M-SLES-NaCl with Surfactant, Salt, Polymer addition order.

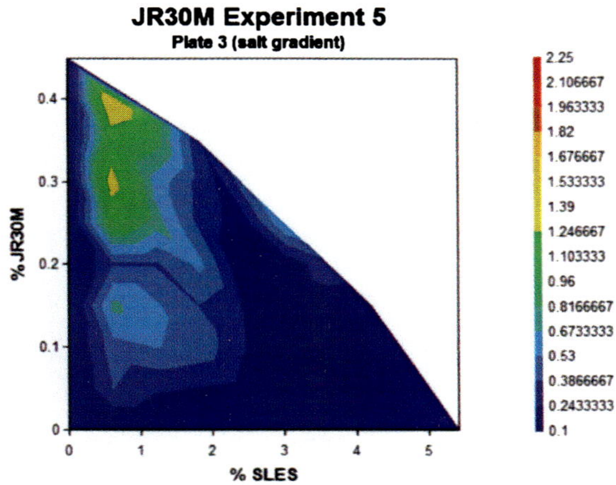

Figure 5.13. Contour phase diagram for JR30M-SLES-NaCl with Polymer, Salt, Surfactant addition order.

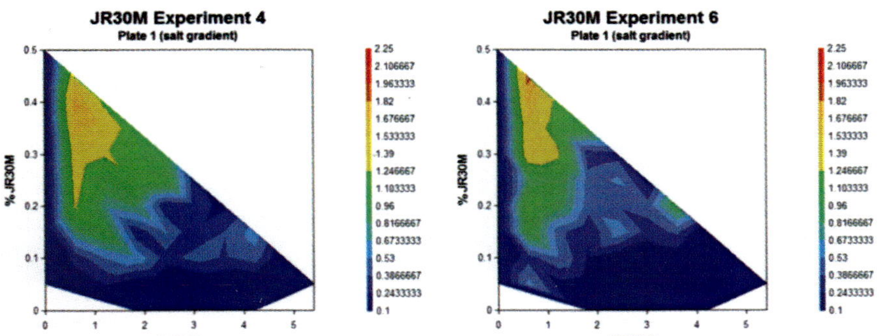

Figure 5.14. Contour phase diagrams for JR30M-SLES-NaCl with (a) Surfactant, Polymer, Salt and (b) Salt, Polymer, Surfactant addition orders.

Chapter 3

Stimuli-Responsive Block Copolymers by RAFT and Their Micellization Behavior

Yuting Li, Brad S. Lokitz, and Charles L. McCormick

Department of Polymer Science, The University of Southern Mississippi, 118 College Drive, Hattiesburg, MS 39406–0076

Here, we review our recent progress in the synthesis of thermally responsive block copolymers by reversible addition-fragmentation chain transfer polymerization. In aqueous solution at room temperature these block copolymer chains exist as unimers but form micelles when the solution temperature is raised above the lower critical solution temperature. Incorporation of an active monomer *N*-acryloxysuccinimide into the triblock copolymer allows for facile formation of uniform shell cross-linked micelles by reaction with ethylenediamine in aqueous media. When cystamine was used as the crosslinker, the cross-linking process is fully reversible: the cleaved SCL micelles can be reformed using cystamine as a thiol-exchange compound. These SCL micelles can be used as potential nanoscale drug delivery carriers, and the rate of drug release can be easily controlled.

Introduction

Controlled/"living" radical polymerization including nitroxide mediated polymerization,[1] atom transfer radical polymerization,[2] and reversible addition-fragmentation chain transfer (RAFT)[3] polymerization have been the research of focus in the last few decades since they are effective for preparing complex architectures such as blocks and stars.[4,5,6] Although each controlled/"living" radical polymerization technique has its own characteristics and advantages, RAFT is arguably the most versatile since it is compatible with most functional monomers under conditions that are similar to conventional free radical polymerization.[7,8] RAFT operates on the principle of degenerative chain transfer. Rizzardo et al. proposed a mechanism for RAFT polymerization as outlined in Scheme 1.[3] The consumption of CTA **1** and reversible fragmentation of species **2** to yield the reinitiating R fragment are often referred to as the "pre-equilibrium" shown in eq b of Scheme 1. Eventually, the "main" equilibrium (eq d) is reached in which active (kinetic) radical chain ends add to monomer or reversibly to dormant chains (termed macro-CTAs). The pseudo-first-order kinetics, the linear evolution of molecular weight with time, narrow molecular weight distributions, and the ability to prepare block copolymers under appropriate conditions attest to significant elimination of chain termination events such as radical coupling and chain transfer common to conventional free radical polymerization shown in **e** of Scheme 1.

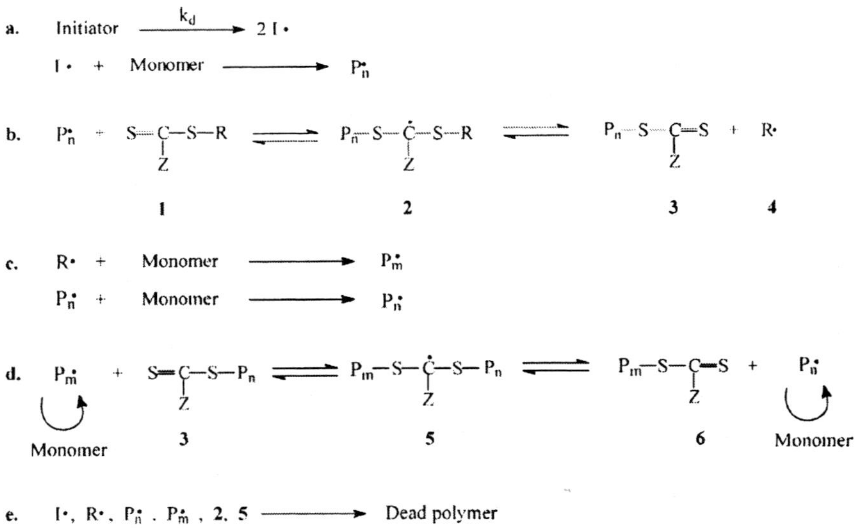

Scheme 1. Proposed RAFT Mechanism. (Adapted with permission from reference 9. Copyright 2002 American Chemical Society.)

By carefully select the appropriate CTA, a wide range of homopolymers, block copolymers and complex architectures can be synthesized using RAFT polymerization technique.[3] Our group has a long-standing interest in the RAFT polymerization and copolymerization of hydrophilic acrylamido monomers[9] including anionic,[10,11] cationic,[12] zwitterionic,[13,14] and neutral derivatives.[15,16,17] Here, we review our recent progress in the synthesis of stimuli-responsive block copolymers and their micellization behaviors. Potential applications for these micelles are also proposed.

Synthesis of thermally-responsive block copolymers by RAFT

Poly(ethylene oxide) (PEO) based macro chain transfer agent was used to mediate the RAFT copolymerization of *N,N*-dimethylacrylamide (DMA) with *N*-acryloxysuccinimide (NAS), forming a PEO-b-(DMA-s-NAS) diblock copolymer, this diblock copolymer was then chain extended with *N*isopropylacrylamide (NIPAM), leads to the formation of PEO-b-(DMA-s-NAS)-b-NIPAM triblock copolymers (scheme 2).[18] All synthetic triblock copolymers have relatively low polydispersities (<1.3), indicating good control of the RAFT copolymerization.

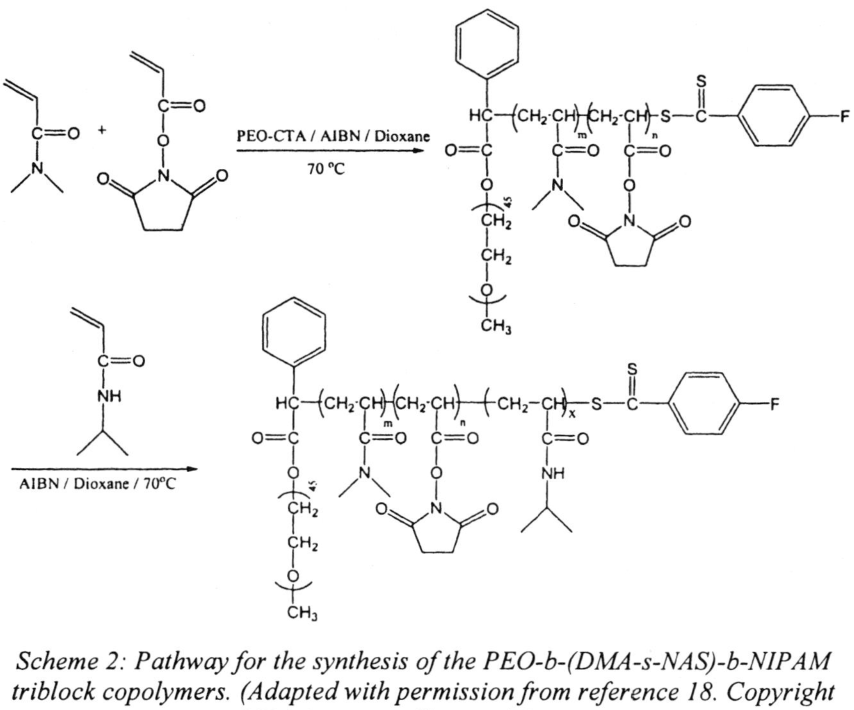

Scheme 2: Pathway for the synthesis of the PEO-b-(DMA-s-NAS)-b-NIPAM triblock copolymers. (Adapted with permission from reference 18. Copyright 2006 American Chemical Society.)

Micellization Behavior

At temperatures lower than the critical temperature, all four triblock copolymers are molecularly dissolved with hydrodynamic diameters (<D_h>) of approximately 7-8 nm. By increasing the solution temperature above 37 °C, micellization occurs for the PEO_{45}-*b*-(DMA_{92}-*s*-NAS_{31})-*b*-$NIPAM_{91}$ and PEO_{45}-*b*-(DMA_{91}-*s*-NAS_{12})-*b*-$NIPAM_{88}$ triblock copolymers. However, PEO_{45}-*b*(DMA_{91}-*s*-NAS_{12})-*b*-$NIPAM_{57}$ and PEO_{45}-*b*-(DMA_{92}-*s*-NAS_{31})-*b*-$NIPAM_{53}$ require higher temperatures (39 °C and 42 °C respectively) to induce micellization (Figure 1). It seems that the micellization temperature is quite sensitive to the molecular weight of the NIPAM block.[19] The higher the molecular weight of the NIPAM block, the lower the micellization temperature.

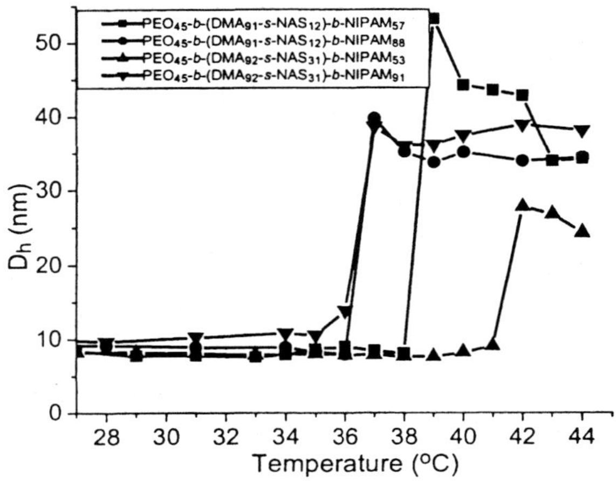

Figure 1. The hydrodynamic diameter vs. Temperature data for the four triblock copolymers in aqueous solution (0.5 % (w/w) concentration). (Adapted with permission from reference 18. Copyright 2006 American Chemical Society.)

Figure 2 shows the NMR spectra recorded for PEO_{45}-*b*-(DMA_{92}-*s*-NAS_{31})-*b*-$NIPAM_{91}$ triblock copolymers at different temperatures. At 25 °C, the copolymer chains are fully solvated and signals expected for each block are visible. At 39 °C, DLS shows that micelles have been formed; although the peaks associated with the NIPAM block have become attenuated, they are still visible in the NMR spectra. Comparison of the NIPAM signal at δ 4.0 with the PEO signal at δ 3.7 indicates a 55% decrease in relative intensity for the former.

This indicates that at 39 °C, the NIPAM cores are still partially hydrophilic. Increasing the solution temperature to 42 °C causes the NIPAM signal to diminish since the NIPAM core becomes increasingly hydrophobic. This is also confirmed by DLS experiments: at 37 °C, the micelle polydispersity is relatively high (0.198), whereas at 39 °C the polydispersity decreases to 0.158. At 42 °C, micelles with uniform size distributions (polydispersity as low as 0.042) can be obtained. The micellar polydispersities are also very sensitive to the NIPAM block length.

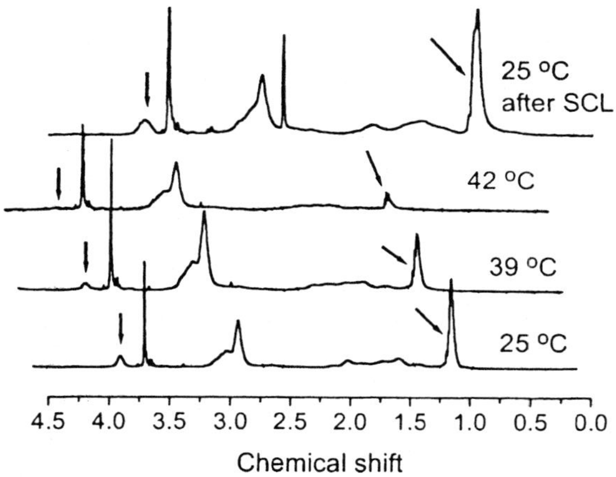

Figure 2. ^{1}H *NMR spectra of the* PEO_{45}*-b-(*DMA_{92}*-s-*NAS_{31}*)-b-*$NIPAM_{91}$ *and the shell cross-linked derivative in* D_2O *at selected temperatures. (Adapted with permission from reference 18. Copyright 2006 American Chemical Society.)*

Shell Cross-Linking

Several crosslinking methods have been reported by other researchers to prepare Shell Cross-Linked (SLC) micelles.[20-27] For our NAS containing thermally responsive micelles, shell cross-linking can be achieved by adding a difunctional primary amine-containing compounds such as ethylenediamine to the micellar solution. Ethylenediamine can react with the NAS residues and hence lead to the cross-linking of the inner shell of the "onion" like micelles. The shell cross-linking process can be easily monitored by ^{1}H NMR. Results show that the maximum degree of cross-linking (>95%) is reached within 2 hours. This is quite fast compared to other methods which normally require several hours to a few days. After crosslinking, lower down the solution temperature will cause the SCL micelles to swell since the NIPAM core become hydrophilic again (Figure 3).

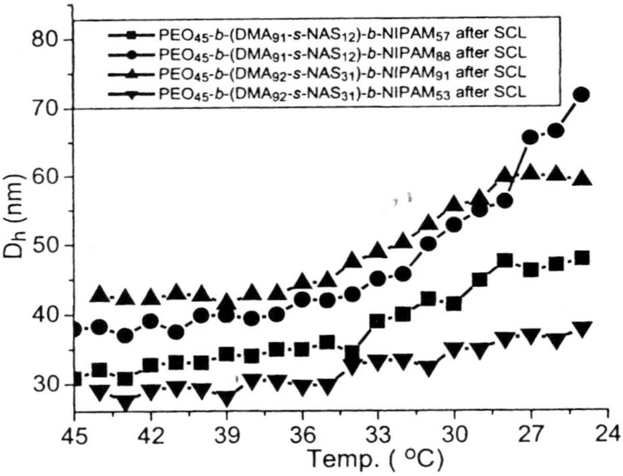

Figure 3. Variation of micelle hydrodynamic diameters with solution temperature after shell cross-linking. (Adapted with permission from reference 18. Copyright 2006 American Chemical Society.)

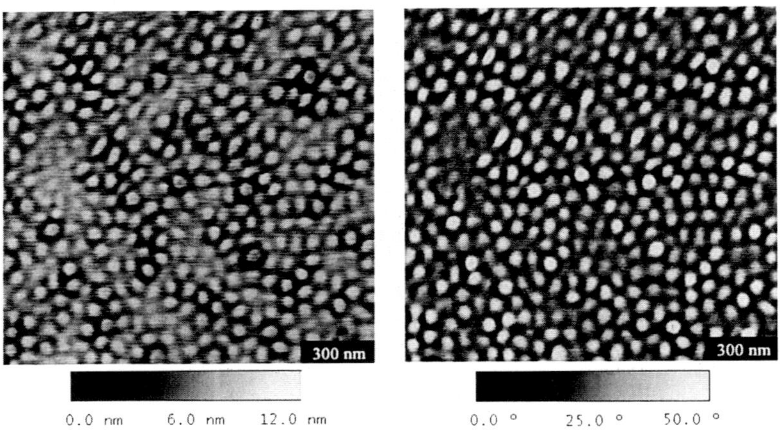

Figure 4. Tapping-mode AFM images of SCL micelles (left: height image; right: phase image). (Adapted with permission from reference 18. Copyright 2006 American Chemical Society.)

Figure 4 shows an AFM image of the SCL micelles prepared from PEO_{45}-*b*-(DMA_{92}-*s*-NAS_{31})-*b*-$NIPAM_{91}$. It can be seen that the micelles are relatively uniform in diameter, approximately 75 nm, slightly larger than in aqueous solution at 25 °C (60 nm). AFM analysis also shows that the height of the SCL micelles is only 8-9 nm (Figure 4, height image), which is significantly less than the micelles diameter. All of these indicate some flattening of the partially swollen (loose) SCL micelles on the AFM substrate as reported by Wooley *et al.*[28,29]

Reversible Shell Cross-Linked Micelles

Cystamine can also be used as cross-linking agent to prepare SCL micelles. At room temperatures the triblock copolymers (PEO_{45}-b-[DMA_{98}-s-NAS_{30}]-b-$NIPAM_{87}$) are molecularly dissolved and have a hydrodynamic diameter of approximately 7-8 nm. Increasing the solution temperature to 37 °C leads to the formation of micelles with a NIPAM core. Compact micelles with a hydrodynamic diameter around 38 nm are formed at 45 °C. The NAS unit in the middle block of the triblock copolymer was then cross-linked using 1 molar equivalent of cystamine. After cross-linking, the solution temperature was lowered to room temperature. DLS results show that the size of the micelles increase to 57 nm. The larger micelle size is due to micellar swelling caused by the core-forming NIPAM block becoming hydrophilic again. Disulfide bonds can be readily broken using either dithiothreitol (DTT)[30,31] or Tris-(2-carboxyethyl)phosphine (TCEP).[32] Here they were used to cleave the disulfide bond in the SCL micelles. DLS results show that the SCL micelles can be fully cleaved because the hydrodynamic diameter decreased from 60 nm to 9-10 nm, which is similar to that of the triblock copolymer precursor. The excess DTT or TCEP and other byproducts can be removed by dialysis against DI water. After cleavage with DTT or TCEP, the disulfide containing crosslinking unit in the middle block of the triblock copolymer is converted to a thiol unit and the triblock copolymer retains its thermal responsiveness. The micelles are reformed when the solution temperature is raised to 45 °C. The micelle size is slightly larger than that of the PEO-*b*-[DMA-*stat*-NAS]-*b*-NIPAM triblock copolymer. These micelles can be re-crosslinked using cystamine. In contrast to the reaction of the cystamine with the NAS unit, the cystamine acts as a thiol exchange reagent allowing the micelles to be crosslinked again. After re-crosslinking, the solution temperature was lowered to room temperature, DLS results shows the particle size is about 75 nm, which means that the micelles have been crosslinked. After reformation by the thiol/disulfide exchange reaction, the chemical structure of the reformed SCL micelles is the same as that before

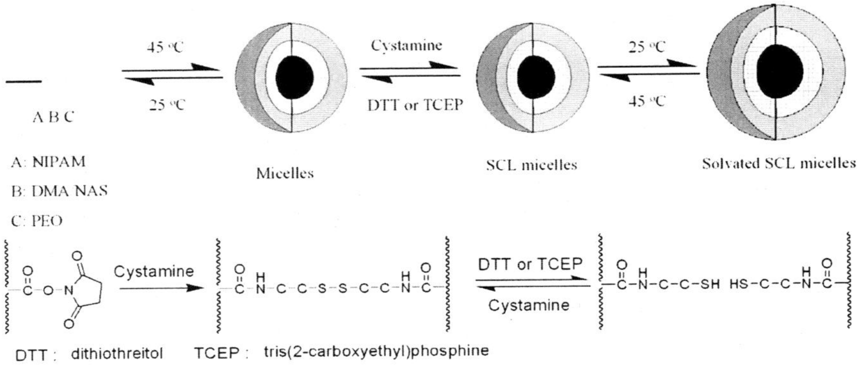

Scheme 3. Schematic illustration of the formation of reversible shell cross linked micelles from PEO-b-(DMA-s-NAS)-b-NIPAM triblock copolymers.

cleavage, therefore, the reformed SCL micelles can be cleaved and reformed repeatedly making the process fully reversible (Scheme 3).

The reversible SCL micelles have potential applications as drug release carrier. Our preliminary results show that a model drug (dipydridamole) can be successfully loaded into the SCL micelles. These drugs were slowly released from the SCL micelles to the environmental PBS buffers. Upon adding DTT to the SCL micelles solution, the drug release rate can be increased since in the presence of DTT, the SCL micelles will be cleaved into single polymer chains. The cleavage rate of the SCL micelles can be controlled by adjusting the DTT concentration or by using a more efficient chemical agent such as TCEP, therefore, the release rate can be easily controlled. The SCL micelles can also be cleaved by using an oxidizing agent such as a peroxide. Our results show that when H_2O_2 was used to cleave the SCL micelles at 60 °C, the process is completed within 10 hours. Since H_2O_2 can be produced in the human body by the immune system, it could be used for the *in-situ* cleavage of the SCL micelles, which would facilitate the elimination of the polymer from body via kidney after drug release.

Conclusion

Well-defined PEO-*b*-(DMA-*s*-NAS)-*b*-NIPAM triblock copolymers were prepared by reversible addition-fragmentation chain transfer polymerization. These thermally-responsive triblock copolymers dissolve molecularly in aqueous solution at room temperature and self-assemble to form micelles when the temperature is raised above the LCST of the NIPAM block. The middle layer of the micelles (DMA-s-NAS block) can be easily cross-linked with ethylenediamine. When cystamine was used as the crosslinker, the cross-linking process is fully reversible: the micelles can be reformed using cystamine as a thiol-exchange compound. These SCL micelles can be used as potential nanoscale drug delivery carriers, and the rate of drug release can be easily controlled.

Acknowledgements

We gratefully acknowledge the financial support provided by the Department of Energy (DE-FC26-01BC15317) and the MRSEC program of the National Science Foundation (DMR-0213883).

References

1. Hawker, C. J.; Bosman, A. W.; Harth, E. *Chem. Rev.* **2001**, *101*, 3661-3688.
2. Matyjaszewski, K.; Xia, J. *Chem. Rev.* **2001**, *101*, 2921-2990.
3. Moad, G.; Rizzardo. E.; Thang, S. H. *Aust. J. Chem.* **2005**, *58,* 379-410
4. M. Kamigaito, T. Ando, M. Sawamoto, *Chem. Rev.* **2001**, *101*, 3689-3745.
5. Matyjaszewski, K., Davis, T. P., Eds. *Handbook of Radical Polymerization*; John Wiley & Sons: Hoboken, 2002.
6. Rizzardo, E.; Chiefari, J.; Mayadunne, R. T. A.; Moad, G.; Thang, S. H. *ACS Symp. Ser.* **2000**, *768*, 278-296.
7. Rizzardo, E.; Chiefari, J.; Chong, Y. K.; Ercole, F.; Krstina, J.; Jeffery, J.; Le T. P. T.; Mayadunne, R. T. A.; Meijs, G. F.; Moad, C. L.; Moad, G.; Thang, S. H. *Macromol. Symp.* **1999**, *143*, 291-307.
8. Moad, G.; Chiefari, J.; Krstina, J.; Postma, A.; Madayunne, R. T. A.; Rizzardo, E.; Thang, S. H. *Polym. Int.* **2000**, *49*, 993-1001.
9. McCormick, C. L.; Lowe, A. B. *Acc. Chem. Res.* **2004**, *37*, 312-325.
10. Sumerlin, B. S.; Donovan, M. S.; Mitsukami, Y.; Lowe, A. B.; McCormick, C. L. *Macromolecules* **2001**, *34*, 6561-6564.
11. Sumerlin, B. S.; Lowe, A. B.; Thomas, D. B.; McCormick, C. L. *Macromolecules* **2003**, *36*, 5982-5987.
12. Vasilieva, Y. A.; Thomas, D. B.; Scales, C. W.; McCormick, C. L. *Macromolecules* **2004**, *37*, 2728-2737.
13. Donovan, M. S.; Lowe, A. B.; Sanford, T. A.; McCormick, C. L. *J. Polym. Sci., Part A*: *Polym. Chem.* **2003**, *41*, 1262-1281.
14. Donovan, M. S.; Sumerlin, B. S.; Lowe, A. B.; McCormick, C. L. *Macromolecules* **2002**, *35*, 8663-8666.
15. Donovan, M. S.; Sanford, T. A.; Lowe, A. B.; Sumerlin, B. S.; Mitsukami, Y.; McCormick, C. L. *Macromolecules* **2002**, *35*, 4570-4572.

16. Thomas, D. B.; Sumerlin, B. S.; Lowe, A. B.; McCormick, C. L. *Macromolecules* **2003**, *36*, 1436-1439.
17. Convertine, A. J.; Ayres, N.; Scales, C. W.; Lowe, A. B.; McCormick, C. L.; Biomacromolecules **2004**, *5*, 1177-1180.
18. Li, Y.; Lokitz, B.; McCormick, C. L. *Macromolecules*, 2006, *39(1)*, 81-89.
19. Xia, Y.; Yin, X.; Burke, N. A. D.; Stover, H. D. H.; *Macromolecules*; **2005,** *38*, 5937-5943.
20. Huang, H.; Kowalewski, T.; Remsen, E. E.; Gertzmann, R.; Wooley, K. L.; *J. Am. Chem. Soc.* **1997**, *119*, 11653-11659.
21. Zhang, Q.; Remsen, E. E.; Wooley, K. L. *J. Am. Chem. Soc.* **2000**, *122,* 3642-3651.
22. Butun, V.; Billingham, N. C.; Armes, S. P. J. Am. Chem. Soc. **1998**, 120, 12135-12136.
23. Butun, V.; Lowe, A. B.; Billingham, N. C.; Armes, S. P. J. Am. Chem. Soc. **1999**, 121, 4288-4289.
24. Liu, S.; Armes, S. P. *J. Am. Chem. Soc.* **2001,** *123*, 9910-9911.
25. Liu, S.; Weaver, J. V. M.; Save, M.; Armes, S. P. *Langmuir* **2002**, *18,* 83508357.
26. Stewart, S.; Liu, G. J. Chem. Mater. **1999**, *11*, 1048-1054.
27. Weaver, J.V.M.; Tang, Y. Q.; Liu, S.; Iddon, P. D.; Grigg, R.; Armes, S. P.; Billingham, N. C.; Hunter, R.; Rannard, S. P. Angewandte Chem. **2004**, *43*, 1389-1392
28. Pan, D.; Turner, J. L.; Wooley, K. L.; *Macromolecules* **2004**, *37*, 7109-7115.
29. Joralemon, M. J.; Smith, N. L.; Holowka, D.; Baird, B.; Wooley, K. L.; *Bioconjugate Chem.* **2005**, *16*, 1246-1256.
30. Tsarevsky, N. V.; Matyjaszewski, K. *Macromolecules*, **2002**, *35*, 9009-9014.
31. Li, Y.; Armes, S. P.; *Macromolecules*, **2005**, *38*, 8155-8162.
32. Cline, D. J.; Redding, S. E.; Brohawn, S. G.; Psathas, J. N.; Schneider, J. P.; Thorpe, C.; *Biochemistry,* **2004**, 43, 15195-15203.

Chapter 4

Advanced Emulsions: Enabling Advanced Emulsion with Microchannel Architecture

Laura Silva[1], Anna Lee Tonkovich[1], Robert Y. Lochhead[2], Dongming Qiu[1], Krissy Pagnatto[1], Paul Neagle[1], Steve Perry[1], and Jan Lerou[1]

[1]Velocys, Inc., 7950 Corporate Boulevard, Plain City, OH 43064
[2]The Institute for Formulation Science, The University of Southern Mississippi, 118 College Drive, #10076, Hattiesburg, MS 39402

An innovative emulsification technology that produces modern emulsions at a reasonable cost has been demonstrated. This process is useful for producing conventional emulsions, and so-called "surfactant free" emulsions, which are stabilized by particles and/or amphipathic polymers. Such emulsions present significant formulation and processing challenges because conventional emulsification techniques rely upon shear force to break up the droplets formed by the discontinuous phase and to transport the emulsifier to the interface. Microchannel emulsification technology is unlike traditional methods that use high-shear forces to form small droplets. This technology, which consists of intervening microchannels with apertured substrates, adds the discontinuous to the continuous phase, one droplet at a time, and provides better control than conventional methods. The droplet size distribution is controlled by adjusting critical process parameters (mixing energy, mixing time), and precise heating and/or cooling.

Emulsions, complex dispersions containing a component that influences fluid-fluid interface stabilization, are essential materials in many chemical-based products, including pharmaceutical, adhesive, food, and personal care products. The characteristics and utility of emulsion products depend on both product formulation and the processing method, which are interdependent. Processing parameters, such as mixing time, mixing energy, and thermal history, impact the final product. For example, the stability of an emulsion typically depends on both its chemistry and the size distribution of the discontinuous phase material, which is typically controlled by the method and energy of phase dispersion. In addition, for an identical chemical formulation, product results can differ when the emulsion is prepared at different volumes because of changes in the processing physics. The formulation of stable emulsions is not a trivial process, and scale-up requires a thorough understanding of both the art and science of emulsified mixtures.

There are three distinct flow conditions: namely laminar flow, turbulent flow, and cavitation. In laminar flow, viscous forces predominate. The flow causes shear or elongation which affects emulsification if a constant flow pattern can be maintained. In turbulent flow, inertial forces predominate. Here the flow conditions vary over small distances and times; therefore modeling of turbulent conditions is difficult and not highly predictive. Cavitation is the most severe condition for emulsification. It occurs when the fluid is accelerated so fast that convection and diffusion cannot keep pace and "holes" are formed in the fluid as the local pressure falls below the vapor pressure of the continuous phase. The vapor bubbles subsequently collapse and cause heavy shock waves that disrupt the droplets.

Emulsions are usually formed under high shear (turbulent or cavitation) conditions in conventional mixers and the agitator configurations can vary from pitched blade turbines, marine impellers, and helical ribbons and sweep blades that favor pumping, to flat blade turbines, pfaudler blades, paddles, and anchor stirrers that favor shear. The complexity and challenges of emulsion formation have also been addressed by specialized equipment such as colloid mills, homogenizers, pohlman whistles, ultrasonic applicators, rotor-stator mixers, and turbines. Turbulent flow or cavitation regimes favor droplet formation but also cause droplet re-coalescence, which limits the lower size of droplets and results in a broad distribution of droplet sizes.

Ideally, emulsions would be formed under laminar flow conditions. This regime enables better control of the process parameters, and improved predictability, which has the potential to take the "guesswork" out of the process, especially scale-up to commercial production. It is appropriate at this juncture to consider what theory can tell us about emulsification in laminar flow conditions.

Laminar Shearing

Laminar shearing causes waves to form at the oil-water interface and, as the amplitude of the waves increases, the liquids "finger" into each other. There are two extreme conditions: 1) where the denser liquid is accelerated into the less dense phase, and 2) where the less dense liquid is accelerated into the denser phase.

Denser Phase Accelerated into Less Dense Phase

If the denser liquid phase is accelerated into the less dense phase, then the frequency of the oscillations = $\sin\Omega^2 t$, where $\Omega = [(\rho 2-\rho 1)gk + \gamma k2]/(\rho 2 - \rho 1)$. And in this equation: $\rho 1$ and $\rho 2$ are the densities of the components, g is the acceleration, γ is the interfacial tension, and $k = \lambda/2\pi$ is the wavelength of the ripples on the interface. In this case, the viscosity will damp the oscillation and emulsion droplet formation will be disfavored.

Less Dense Phase into Denser Phase

On the other hand, if the less dense fluid is accelerated into the denser fluid, the amplitude of the disturbances increase exponentially: The rate of growth reaches a maximum where: $\Omega^2 = \{2/3[-(\rho_2-\rho_1)g]^{3/2}\}/[(\rho_2 - \rho_1)(3\gamma)^{1/2}]$, where γ is the interfacial tension and in this case, the less dense liquid moves at a velocity given by $\{[(\rho_2 - \rho_1)/(\rho_2 + \rho_1)]g\lambda\}^{1/2}$ and the denser fluid moves at a velocity given by $1/2[(\rho_2-\rho_1)/(\rho_2 + \rho_1)] + 2$.

This causes the less dense liquid finger to have a blunt tip and the denser fluid fingers to have a spiked tip. The spiked tip stretches into a narrow liquid fiber and Rayleigh-Taylor instability causes this fiber to break into droplets. The breakup into droplets is intensified by Kelvin-Helmholtz instability because the fluids are moving at different speeds.

Understanding the mechanisms of droplet formulation in laminar flow emulsification enables fluid modeling and equipment design of systems which can realize significant improvements in controlling emulsion product characteristics.

Technical Approach

Laminar flow emulsification is possible with microchannel processing. Microchannel processing for chemical process intensification has seen significant worldwide development over the last decade [1]. This technology consists of narrow channels, characterized by having at least one dimension that is typically less than 1 millimeter (mm). Short transport distances from the center of the fluid flow to the channel wall enable fast diffusion for heat and mass transfer. Scale-up is accomplished by parallel processing through increased numbers of microchannels. The reported microchannel technology is based on an architecture that stacks layers of thin sheets of material with stamped or etched features to create flow paths. This architecture offers a wide variety of flow configurations, including very close integration of process and heat transfer fluids. In addition, multiple operations, such as mixing and heat transfer, can be close-coupled in integrated, compact hardware units.

Unlike conventional emulsification methods that involve forces acting on the entire bulk volume of fluid, microchannel processing applies shear forces locally, thereby reducing the mixing energy requirements. This also makes possible the use of shear sensitive chemicals. Only one other emulsification technology can boast similar benefits – Membranes. However, membrane emulsification does not offer all the advantages of microchannel hardware.

Membrane emulsification methods involve introducing a dispersed phase through a porous material into a continuous phase, typically through a cylindrical glass or ceramic tube. Thorough reviews of membrane emulsification can be found in the literature (*7,8*). Membrane emulsification has the potential for reduced energy requirements and the opportunity to tailor the droplet size of the dispersed phase (*5,7*). While cylindrical membrane emulsifiers are capable of producing mixtures with narrow droplet size ranges using lower energy input, they are limited in their ability to achieve high loadings, especially for submicron droplets. Typically membranes can only produce droplets that are 2 to 10 times the average membrane pore size (*7,8*). Other methods have also been developed, such as the co-flow single droplet method (*9*), which also showed a droplet size to pore diameter ratio of 2 to10, and a micromixing based method (*10*) for which the droplet size produced (20 to 30 microns) was not competitive with methods enabling submicron droplet formation.

Microchannel Emulsification

The microchannel emulsification concept described in this paper (*11*), and seen in Figure 1, is similar to conventional cylindrical membrane systems in that an apertured substrate is used to introduce the dispersed phase as the continuous phase sweeps the droplets off the surface via cross-flow shear. However, because the microchannel emulsifier uses a planar architecture, it offers several advantages over other membrane systems. First, heat exchange channels are more easily integrated with process (emulsification) layers for tighter control of temperature and the possibility of very fast heating or cooling rates before and/or after emulsification. Second, in the planar architecture, a narrow gap for continuous flow (e.g. 0.25 – 2 mm) along the membrane interface can more easily be maintained than in tubular membrane emulsifiers. Also beneficial is the enhanced ability to control the kinetics of emulsification.

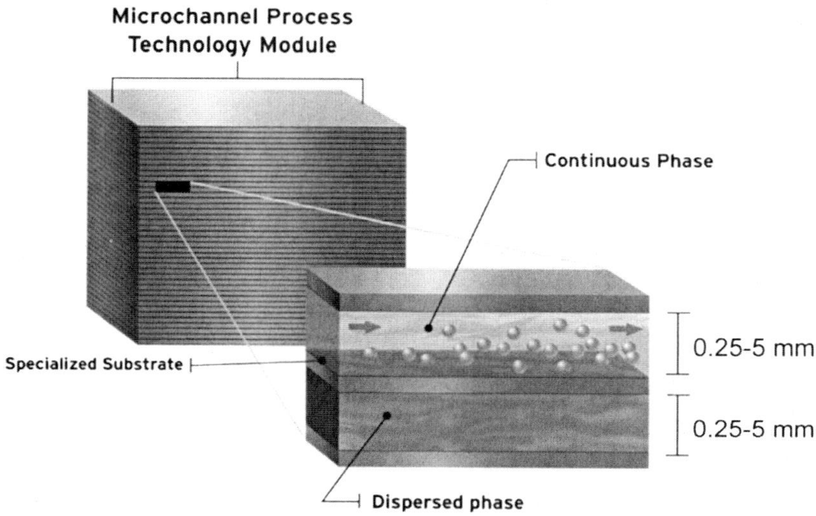

Figure 1. Schematic of a multi-channel or high throughput microchannel substrate emulsification technology

Heat Exchange Integration

These interleaved sheets can be stacked in multiple layers, allowing designs that integrate heat exchange channels. The ability to tailor temperature profile (i.e. rapid quench to freeze-in small droplets or particles which are unstable at emulsification temperatures) provides a method to producing stable emulsified products. The graph in Figure 2, shows that interleaved heat exchange channels can rapidly cool an emulsified mixture from 70°C to 40°C in a span of 0.25 meters or 50 milliseconds. This quenching occurred with a flow of 1 LPM per channel of a shear thinning non-Newtonian fluid with a viscosity of 100 cP.

Narrow Gaps

The narrow gaps possible with microchannel architecture have at least two advantages: the achievable local shear stress at the substrate interface is higher, while the shear stress in the bulk fluid remains very low, and the substrate interface area per volume of continuous phase channel is higher, allowing higher loadings per pass. This effect is depicted in Figure 3, where a decreasing microchannel gap (δ) increases the velocity gradient at the wall. This, in turn, can translate into increased wall shear stress. The component of shear stress

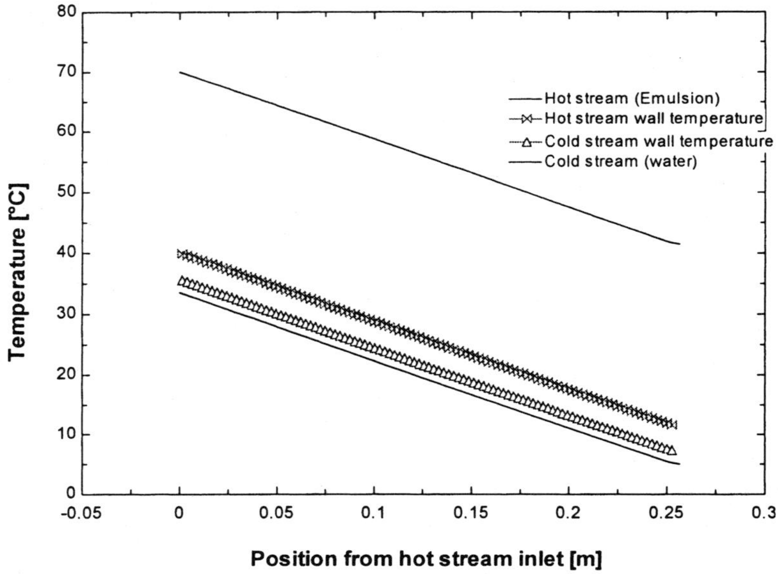

Figure 2. Integrated cooling channels can rapidly quench emulsified products

most relevant to emulsion droplet formation is the component parallel to the substrate surface that causes successive parallel layers of liquid flow to move in their own planes relative to each other and provides sufficient force to detach small emulsion droplets. This component of shear stress may be calculated as shown in Equation 1,

$$\tau_w \equiv \tau_{yx} = -\mu(T,\dot{\gamma})\frac{\partial u_x}{\partial y} \quad (1)$$

where μ_x denotes the velocity component in the x (axial) direction of continuous phase flow and y is the dimension normal to the substrate surface into the channel gap. The fluid shear in the bulk flow is very low, opening the possibility of producing shear sensitive materials and narrow droplet size distributions.

Table I compares the typical shear rates achievable with the microchannel emulsification technology at the wall and in the bulk with other technologies, most of which expose the entire bulk to the same shear rate. Only the rotor/stator and homogenizer have similar shear rates, but this shear is imparted to the entire fluid.

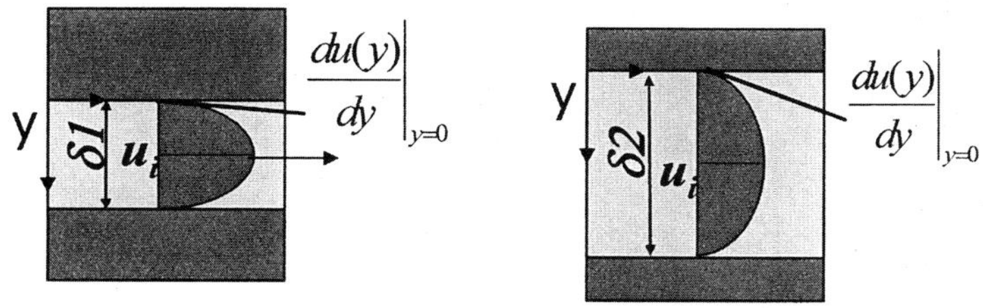

Figure 3. Schematic showing increase in shear force as channel gap decreases and the velocity of the continuous fluid increases. (Reproduced with permission from Cosmetics and Toiletries ***August 2005,*** 120 (8). *Copyright 2005.)*

Table I. Technology Comparison of Shear Rate Range

Technology	*Shear rate range*
Impeller	10 -1,000
Rotor/stator and homogenizer	20,000- 100,000
Microchannel-wall	10,000-120,000+
Microchannel-bulk	up to 5,000

NOTE: Units are inverse seconds (1/s)

SOURCE: Reproduced with permission from *Cosmetics and Toiletries* **August 2005**, *120 (8)*. Copyright 2005.

Kinetics of Emulsification

One final advantage of the planar microchannel emulsification technology is control of the kinetics of emulsification. Effective emulsification requires the emulsifier to adsorb at the oil- water interface as it is formed. For example, if the emulsifier is present in a micellar form, the micelles act as reservoirs that supply emulsifiers to the expanding oil-water interface. If micellar dissociation and diffusion of emulsifier to the interface is too slow, the droplets will re-coalesce faster than they can be stabilized. Lamellar liquid crystalline micellar phases are desired as an emulsion stabilizer, but the kinetics of micellar dissociation for this type of micelle is exceedingly slow. Microchannel emulsification offers the emulsion processor the opportunity to match the rate of drop formation with the rate of demicelization and diffusion to the interface and by this means unprecedented stable and efficacious emulsions will become possible.

As shown in Figure 4, the apertured substrate is disposed between intervening channels of continuous-phase and dispersed-phase feeds. A simple three-layer example of using the microchannel and substrate technology to achieve reduced emulsion droplet size has been demonstrated for a variety of formulations.

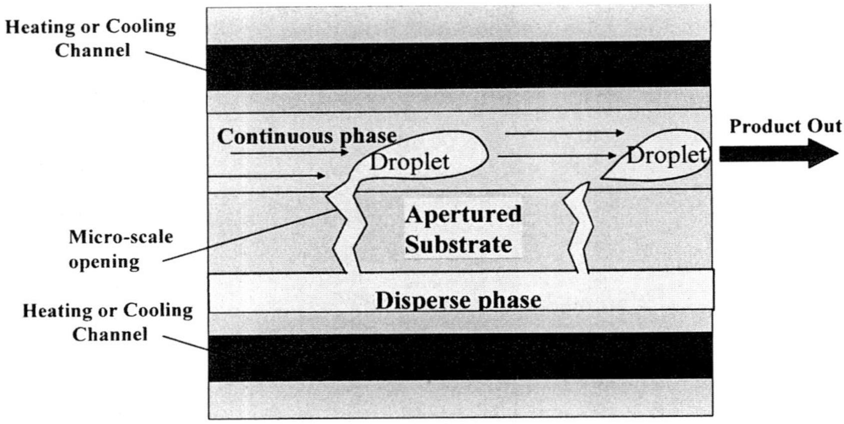

Figure 4. Schematic of channel configuration in microchannel emulsifier technology concept. (Reproduced with permission from Cosmetics and Toiletries ***August 2005***, 120 (8). *Copyright 2005.)*

Experimental System

The experimental system used to form the oil in water emulsions was a multiple-pass recycle system to which the dispersed phase was gradually added. A recycle system was used to achieve high dispersed-phase loadings in a small prototype emulsifier. The system included a dispersed-phase pump (Model QG6, FluidMetering, Inc.), a continuous-phase pump (Micropump Model GJ-N25, JF18A, Cole Parmer), a dispersed-phase reservoir, a continuous-phase reservoir, and the microchannel emulsifier. Flow rates were measured by rotameter (Cole Parmer), calibrated for a range of compositions. The emulsion formulation chosen for this example required processing at slightly elevated temperatures to prevent solidification. Electrical resistance heating tapes with individual control thermocouples and a uniform layer of thermal insulation were used to maintain the stainless steel transfer lines and all hardware components at about 75°C.

The microchannel emulsifier (Figure 5) was designed to provide a high aspect ratio rectangular gap (0.5 mm by 12.7 mm). The small dimension was selected to provide for sufficient fluid shear during droplet formation. The large dimension was chosen to provide for sufficient capacity for continuous phase flow. The device was made in a series of layers that were gasketed and bolted together to form a leak-tight operating device.

The exposed substrate surface area in the emulsifier was 12.7 mm by 12.7 mm. The substrate used in the emulsifier was a 1-mm thick MOTT

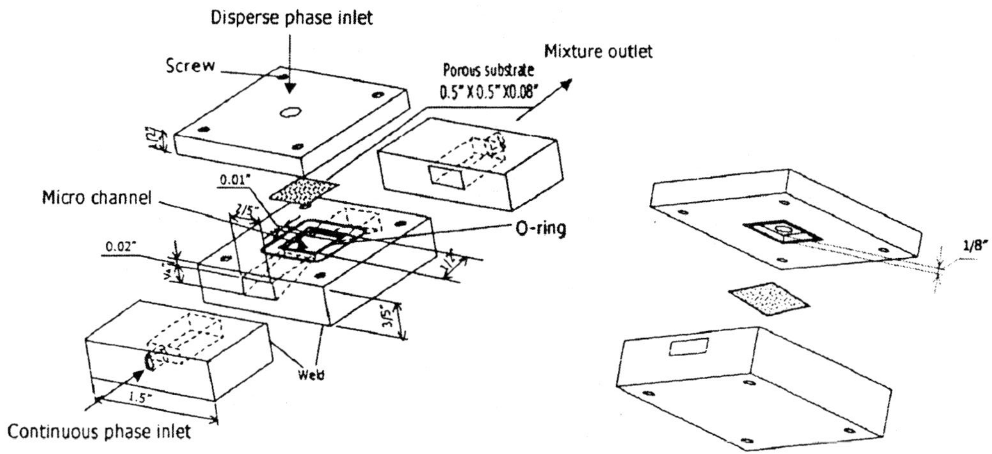

Figure 5. Schematic of the microchannel emulsifier prototype

(Metallurgical Corp., Farmington, CT) stainless steel sintered metal plate with a nominal pore size of 0.5 micron. The MOTT substrate was cleaned prior to use by sonication in hexane, nitric acid, de-ionized water, and acetone or isopropanol and heat treated in air at 650°C for 100 hours. Heat-up at 3.5°C per minute in a hydrogen water mixture, followed by a nitrogen purge. It was estimated by water permeability tests that the nominal pore size was reduced from 0.5 micron to 0.44 micron after heat treatment.

The dispersed phase flowed through the pores of the substrate and was continuously sheared to form small droplets by the cross-flowing continuous phase. Rubber gaskets 0.13 mm thick were used to seal the perimeter of either side of the porous substrate to prevent bypass of the dispersed-phase flow. A dispersed-phase header upstream of the porous substrate was provided for good flow distribution across the substrate.

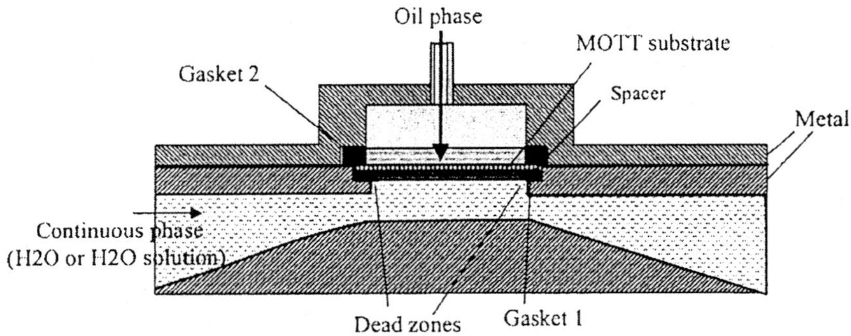

Figure 6. Close-up of cross-section of the microchannel emulsifier prototype. (Reproduced with permission from Cosmetics and Toiletries ***August 2005****,* 120 (8)*. Copyright 2005.)*

The emulsifier was operated to form a moisturizing lotion (an oil-in-water emulsion) having the formula given in Tables II and III. The dispersed (or oil) phase was prepared by mixing all the components except the triethanolamine in a beaker and heating it to 75°C, then adding the triethanolamine. The continuous (or aqueous) phase was prepared by first dispersing Carbopol 934 in the water and heating it to 75°C before adding the remaining ingredients. The dispersed/oil phase flow rate was varied between 2.5 and 10 milliliters per minute (ml/min), and the continuous-phase flow rate was varied from 0.5 to 2.0 liters per minute. The emulsion was continuously recycled through the emulsifier while continuously adding additional oil phase until the target weight loading of 12.7 wt.% oil phase was reached. The product emulsion was quickly cooled to by placing the reservoir in an ice water bath in order to control the cooling for the final product. Droplet size was assessed by a Malvern Particle Size Analyzer (PSA) and optical microscopy when possible. The pH of the final product was 6.5.

Table II. Composition of Continuous/Aqueous phase (87.3% by wt.)

Component	*Parts by weight*
Water	82.9
Carbopol 934 (BF Goodrich/Harris and Ford)	0.20
Na2 EDTA (Dow Chemical Company)	0.05
Glycerine USP (Humco)	4.0

Table III. Composition of Dispersed/oil phase (12.7 % by wt.)

Component	*Parts by weight*
Stearic Acid	2.0
Cetyl Alcohol	0.5
Glyceryl Monostearate	0.2
Ethylene Glycol Monostearate	0.3
Propylparaben	0.1
Methylparaben	0.2
Mineral Oil	7.0
Silicone Fluid DC200 (Dow Corning)	1.0
Tween 20 (Uniqema Americas, ICI)	0.5
Triethanolamine	0.9

Results and Discussion

The moisturizing lotion produced in the microchannel emulsifier as tested with apertured substrates with a nominal 0.1 and 0.5-micron pore size produced droplet sizes ranging from 0.4 to 4 microns, as shown in Figures 7 and 8. Oil phase pressure drop (from oil inlet to product outlet) was 70 to 140 kPa, depending on the flow conditions. The results are consistent with membrane emulsification findings where the average droplet diameter is proportional to the average membrane pore size. These results suggest that droplet sizes in the described microchannel emulsifier can (given the right process conditions) more closely approach the substrate pore size than in tubular membrane technology (a ratio as low as 1.4 to 1 as compared to reported factor greater than 3 for membrane emulsifiers).

The same emulsion formulation was also prepared using a batch mixer and compared for quality and consistency. The formed droplets from the batch mixer were bi-modal, with some large droplets exceeding 10 microns on average and many droplets between 2 and 3 microns. The droplet size distribution graph in Figure 9 shows the ability to tailor emulsion particle sizes by changing process conditions.

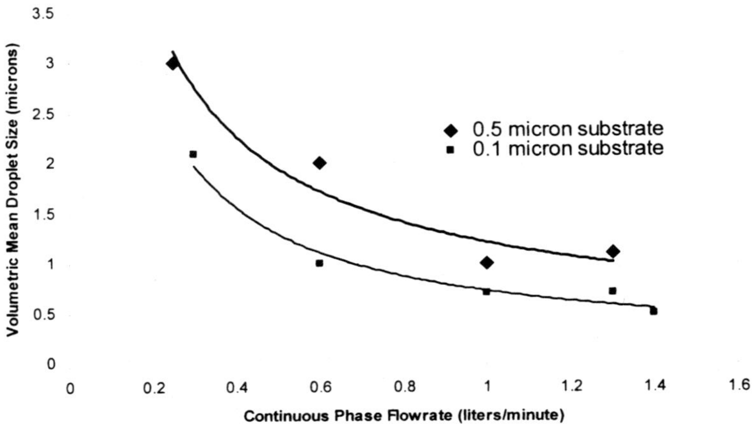

Figure 7. Emulsion mean droplet size as a function to continuous phase flowrate and a function of the substrate pore size for a constant oil flowrate of 10 mL/min (Reproduced with permission from Cosmetics and Toiletries *Copyright 2005.) (See page 2 of color inserts.)*

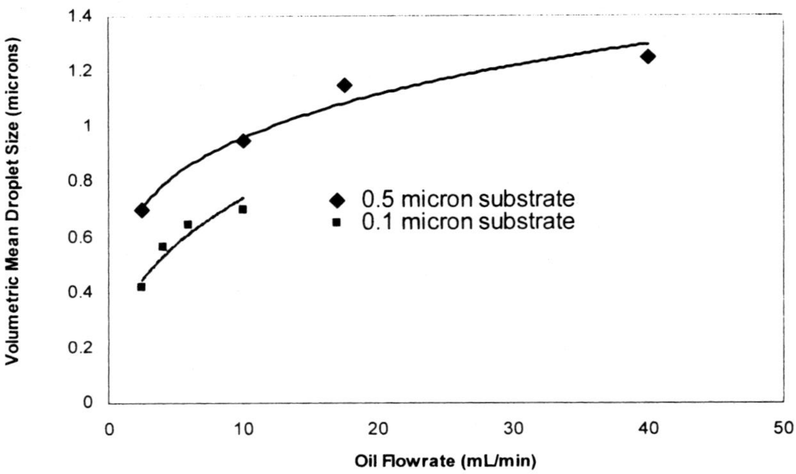

Figure 8. Emulsion mean droplet size as a function of the oil phase flowrate and a function of the substrate pore size for a constant water flowrate of 1 LPM (Reproduced with permission from Cosmetics and Toiletries *Copyright 2005.) (See page 2 of color inserts.)*

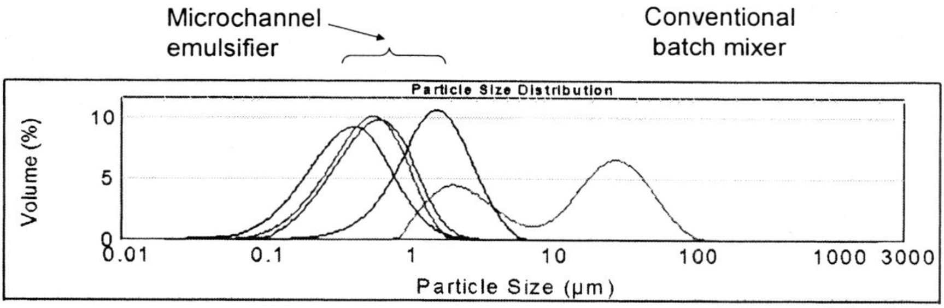

Figure 9. Droplet size distribution can be tailored using microchannel emulsifiers

Conclusions

The microchannel emulsification technology offers advantages for reducing mean droplet size, reducing bulk fluid shear, and improving ease of integration and scale-up. The technology has demonstrated this advantage for oil in water applications, such as hand lotion with a mean droplet less than 1 micron. The mean droplet size was further found to be a function of the processing conditions. Additional work is ongoing to extend the technology to other emulsion formulation systems and address the many scale-up issues required for commercialization.

Microchannel emulsification offers the opportunity to produce emulsions drop by drop in laminar flow and by controlling the pressure drop across the substrate, it will become possible to reliably accelerate the less dense phase into the denser phase. By this means, the emulsification process can be controlled and theoretical predictions and reliable modeling for scale-up will become possible.

Furthermore, microchannel technology can screen an expanded suite of processing conditions to allow previously impossible product formulations with shear-sensitive materials and controlled droplet size distribution.

References

1. Volker H, Hardt S, Löwe H. Chemical Micro Process Engineering: Fundamentals, Modelling and Reactions. 2004. Wiley-VCH Verlag GmbH & Co. KGaA.
2. Tonkovich AY, Perry S, Wang Y, Qiu D, LaPlante T, Rogers WA. Microchannel Process Technology For Compact Methane Steam Reforming. *Chemical Engineering Science*, 59 (2004), 4819-4824.

3. Mazanec TJ, Perry ST, Tonkovich AY, Wang Y. Microchannel Gas-to-Liquids Conversion – Shrinking the Problem. Presented at CatCon, Houston TX, May 2003
4. Canselier JP, Delmas H, Wilhelm AM, Abismail B. Ultrasound emulsification - An overview. *Journal of Dispersion Science and Technology*, v 23, n 1-3, 2002, p 333-349
5. Schubert, H. Mechanical Emulsification – New Developments and Trends. Fall AIChE Meeting held in Los Angeles, November 12-17, 2000 (Surface Phenomena Effects in Liquid/Liquid Mixing). http://www.mixing.net/namf/conferences/2000-AIChE-Fall/folien.pdf.
6. http://www.imm-mainz.de/v0/objekte/pdf/star_laminator_e.pdf
7. Charcosset C, Limayem I, Fessi H. The Membrane Emulsification Process- A Review. *Journal of Chemical Technology*, v 79, 2004, p 333-349
8. Joscelyne SM, Tragardh G. Membrane Emusifications- A Literature Review. *Journal of Membrane Science*, v 169, n 1, Apr 2000, p 107-117
9. Umbanhowar PB, Prasad V, Weitz DA. Monodisperse Emulsion Generation Via Drop Break Off In A Coflowing Stream. Langmuir, v 16, 2000, p 347-351
10. Tong J, Nakajima M, Nabetani H. Preparation Of Phospholipids Oil-In-Water Microspheres By Microchannel Emulsification Technique. *Eur. J. Lipid Sci. Technol.* v 104, 2002, p 216-221
11. Qiu D, Tonkovich AY, Silva L, Long RQ, Yang BL, Trenkamp KM. Process For Forming An Emulsion Using Microchannel Process Technology. US2004/0234566 A1 patent application

Chapter 5

Investigation of the Structure of Polyelectrolyte-Based Complex Coacervates and the Effects of Electrolyte Order of Addition

Lisa R. Huisinga and Robert Y. Lochhead

The Institute for Formulation Science, School of Polymers and High Performance Materials, The University of Southern Mississippi, 118 College Drive, Hattiesburg, MS 39406–0076

High-throughput screening formulation methods were used to investigate structure-property relationships of oppositely charged polyelectrolytes and surfactants in the semi-dilute and concentrated surfactant regimes. These methods were also employed to investigate the mechanisms of coacervate formation in the presence of electrolyte, specifically variation in addition order. Using synthetic polymers it was determined that the positioning of the cationic group along the backbone chain can impact the amount of coacervate formed. Within the cationic cellulosic Polyquaternium-10 polymer series, an increase in molecular weight and/or charge density provided an overall increased amount of coacervate. Also, it was determined that, depending on the flexibility of the polymer structure, the mechanisms of coacervation in the presence of added electrolyte can vary as a function of addition order of the materials.

Introduction

Complex coacervates are useful in a variety of applications, including controlled release drug delivery and conditioning agents in personal care products. Complex coacervate formation is a result of the interactions of two macromolecules, or a macromolecule and an association colloid, each carrying an opposite charge. The interaction of these molecules is promoted by an ion-exchange process where an entropy increase is observed when counterions are released during the ion-exchange process. As a compliment to these electrostatic interactions, in the case of a macromolecule complexing with an association colloid, the hydrophobic tail groups of the surfactant associate and cause the polymer structure to lose conformational entropy. This collapse of the polymer causes a phase separation. The separated phase that is rich in the polymer/surfactant complex, but poor in water, is known as the coacervate (*1-5*). The mechanism of coacervate formation has been widely studied in the dilute regime, near the critical micelle concentration (*1,6-15*). However, little research has focused on polyelectrolyte-surfactant interactions in the semi-dilute and concentrated surfactant regime (*16*). Since many commercial products require coacervates that have been formed in the concentrated polymer-surfactant regime, there is a need to understand this interaction mechanism.

It is also of interest to understand the effects of electrolytes on polymer-surfactant interactions. The presence of an electrolyte in an oppositely-charged polymer-surfactant system should have significant effects on the coacervation process because coacervation is governed by both ion-exchange and hydrophobic associations. Electrolytes in solution could compete with the ion-ion interaction of polymer and oppositely charged surfactants by shielding the polymer from the surfactant. This would limit ion-exchange and ultimately decrease the amount of coacervate formed (*15*). The presence of ions in solution can also affect ionic micelle structure by changing the position of the Donnan Equilibrium at the micelle surface, resulting in a reduction of the effective Bjerrum length. This will cause a decrease in micelle curvature, an increase in micelle size and a change in micelle shape from spherical to cylindrical, to planar, and eventually to inverse micelles. Water–structuring electrolytes in solution will also enhance the hydrophobic interaction between the surfactant molecules. In light of these hypotheses, one goal of this research was to investigate the effect of salt on oppositely-charged polymer-surfactant systems.

A comprehensive investigation of complex coacervation in the semi-dilute and concentrated regimes requires examination of an immense number of samples to cover a range of polymer, surfactant, and electrolyte concentrations. In order to achieve these goals within a reasonable time frame, it is desirable to employ high-throughput screening techniques. Techniques for high-throughput formulation do not exist in the literature, so their development was the initial step in the work presented here. Using a liquid handling system for sample preparation, we are able to analyze nearly 1000 samples per day, making the above goals of understanding electrolyte effects and coacervate structure-property relationships attainable. The phase diagrams that are produced using these high-throughput screening techniques are used as a guide to exact compositions that can be investigated using more powerful analyses, such as small-angle neutron scattering (SANS), rheology measurements, and electron microscopy. Combining the knowledge gained from the phase diagrams and the detailed analyses, a predictive model can be developed for structure-function relationships of complex coacervates.

Experimental

Materials

Poly (4-vinyl pyridine) and poly (2-vinyl pyridine) were synthesized via Reverse-Addition Fragmentation Chain Transfer (RAFT) polymerization and characterized in the McCormick research group according to previously published methods (*17*). The chain transfer agents used for poly (4-vinyl pyridine) and poly (2-vinyl pyridine) were 4-cyanopentanoic acid dithiobenzoate (CTP) and 2-dodecylsulfanylthiocarbonylsulfanyl-2-methyl propionic acid (DMP), respectively. The molecular weight (M_n) and polydispersity index (PDI) were determined for both polymers using Size Exclusion Chromatography (SEC). The M_n of poly (4-vinyl pyridine) was 6,717 with a PDI of 1.037. The M_n of poly (2-vinyl pyridine) was 25,762 with a PDI of 1.055. Commercial polyquaternium-10 polymers of varying molecular weights and charge densities were obtained from Amerchol Corp. and used as received. Table I lists the molecular weights and charge densities (% Nitrogen) of each polymer studied. The molecular weight values listed in Table I were provided by Amerchol. Svensson and coworkers (*16*) evaluated the molecular weight of UCARE™ polymer JR400 and determined it to be ~500,000 g/mol, thus the

Table I. Cationic cellulosic polymers investigated for molecular weight and charge density effects.

Polymer Trade Name	Reported Molecular Weight	Reported Percent Nitrogen
UCARE™ Polymer LK	350,000	0.50
UCARE™ Polymer LR400	350,000	0.95
UCARE™ Polymer JR125	350,000	1.85
UCARE™ Polymer JR400	450,000	1.85
UCARE™ Polymer JR30M	2,000,000	1.85
UCARE™ Polymer LR30M	1,800,000	0.95

molecular weights supplied for the polymers in Table I were accepted. The percent nitrogen values were also provided by Amerchol Corp. (*18*).

The anionic surfactant sodium lauryl ether (3EO) sulfate, 28.82% actives, was obtained from Stepan Co. and used as received. The anionic surfactant sodium dodecylbenzene sulfonate, 23% actives, was obtained from Rhodia Inc. and used as received. Sodium chloride (NaCl), A.C.S. certified grade, was used as received from Fisher Scientific. Distilled, deionized water was used in all samples.

Sample Preparation

The liquid handler used for high-throughput formulation has a viscosity limitation of 1000cps and cannot distribute solids. To overcome these constraints, pre-mixes of the polymer and surfactant were made. The poly (vinyl pyridine) pre-mixes were made by dispersing the polymer powder in water and adjusting the pH below 4 using 1.0N HCl, until the polymer was dissolved. The cellulosic polymer pre-mixes were made by slowly adding polymer powder to water under constant agitation at 45-50°C. These pre-mixes were preserved using 0.003% actives Kathon CG. Both surfactants were supplied as a viscous liquid and were diluted with water to make pre-mixes of a given % actives. Pre-mixes of sodium chloride were made by dissolving NaCl solids in water.

Formulation samples for phase diagram construction were prepared using a Beckman Coulter Biomek FX Laboratory Automation Workstation. Programs were designed to create approximately 200 samples in less than one hour, with repetitions included to ensure accuracy. 96-well plates with glass vials were used for sample vessels. Once all materials were delivered to the 96-well plates, the samples were mixed using a 96-well plate attachment on a Scientific Industries Vortex Genie 2.

Phase Separation Analysis

Phase separation is used to indicate the relative amount of coacervate formed in each sample. UV-Visible spectrophotometric absorbance measurements were taken on each well in the 96-well plate using a Tecan Safire Multifunction Multiplate Reader. When a new system is first subjected to this analysis, an absorbance scan from 230-990nm (20nm step) is performed on wells that show no, little and high amounts of visible phase separation in order to determine the most desirable fixed wavelength for absorbance readings. For the cationic polymer systems described here, the desired wavelength is 410nm. Using this fixed wavelength, absorbance readings were performed on all wells in the 96-well plate immediately after mixing. The average plate analysis time is less than 2 minutes. The absorbance units are plotted as a function of %

actives polymer and % actives surfactant. A sample contour phase diagram is shown in Figure 1.

A color gradient is used to represent the amount of UV-Vis absorbance in the contour phase diagrams. The absorbance units are plotted such that blue areas indicate no phase separation and red areas indicate the highest amount of phase separation observed experimentally for that particular class of polymers (e.g., the maximum absorbance unit (red) differs between the poly (vinyl pyridines) and the cellulosic polymers). The phase separation color gradient is correlated to the amount of coacervate, where high phase separation (red) indicates a high amount of coacervation. Using these color gradient maps, trends in coacervation as a function of polymer property are easily visible.

Since this is a novel coacervate screening method, the reproducibility of these phase diagrams was investigated. Red and green color coded diagrams covering the same compositional range as the contour phase diagrams are utilized to represent the reproducibility. The work presented here on the poly (vinyl pyridines) was performed with a different program on the liquid handling instrument than the cellulosic polymer work. For the synthetic systems, repeat compositions were not incorporated into the same 96-well plate, but a complete duplicate experiment was performed so that all compositions have two repetitions. For all compositions the absolute value of the difference between the two absorbance readings was calculated. The results of this calculation are

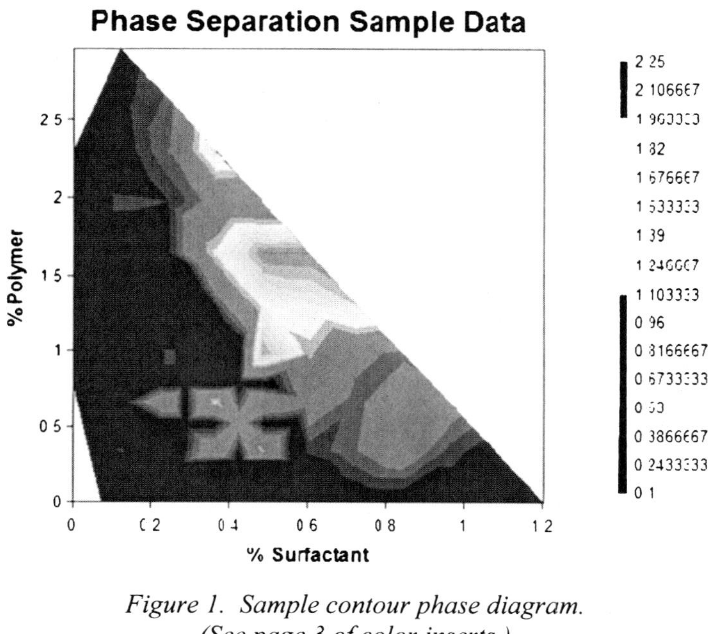

Figure 1. Sample contour phase diagram.
(See page 3 of color inserts.)

plotted as a function of the composition. The degree of difference between repetitions as a function of the total range of absorbance readings is presented via color coding, where all compositions in green are reproducible within 20% of the range and all compositions in red vary more than 20%. For screening purposes, reproducibility within 20% is deemed adequate to determine gross differences between the systems, and most showed good reproducibility at this level. Typical reproducibilities for the synthetic polymer method are shown in Figure 2. The diagrams on the left are phase separation contour phase diagrams

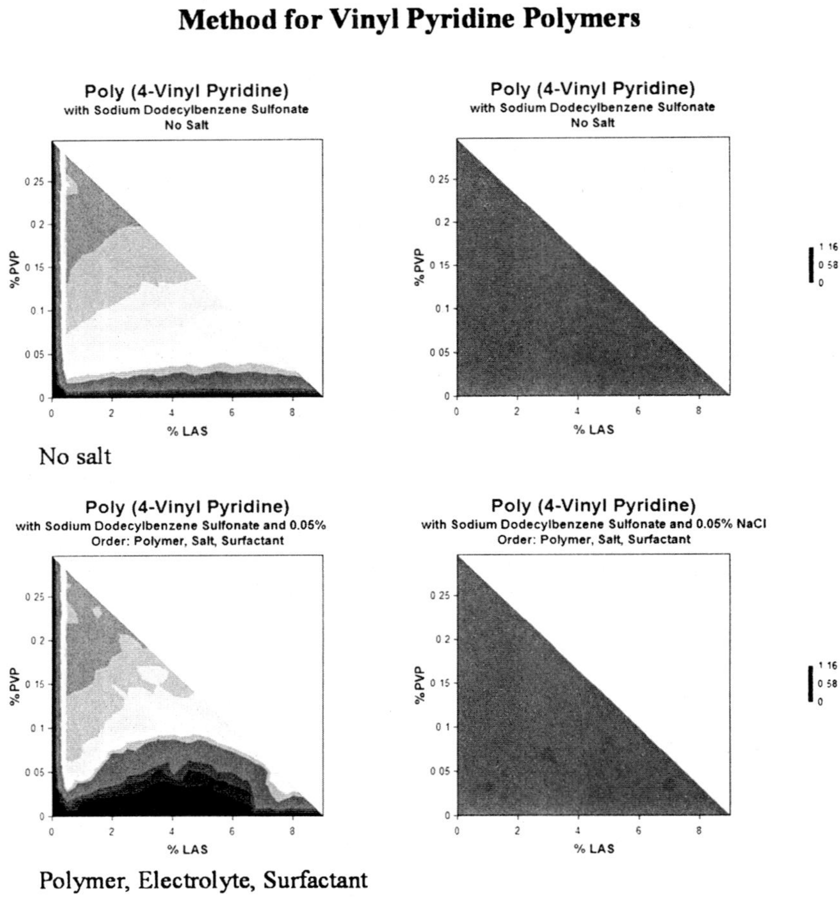

Figure 2. Reproducibility of high-throughput screening methods for synthetic polymer-surfactant interaction contour phase diagrams.
(See page 3 and 4 of color inserts.)

and those on the right are the corresponding reproducibility diagrams. Typical values for experiments with and without the incorporation of salt are presented.

For the cellulosic systems, repeat compositions were incorporated into the same 96-well plate so that some compositions had two or more repetitions. For compositions with only two repetitions, the absolute value of the difference between the absorbance readings was calculated. For compositions having more than two repetitions the standard deviation of absorbance readings was determined. The results of these calculations are plotted as a function of the composition. The degree of difference between repetitions as a function of the total range of absorbance readings is presented via color coding, where all compositions in green are reproducible within 20% of the range and all compositions in red vary more than 20%. Typical reproducibilities for the cellulosic polymer method are shown in Figure 3. The diagrams on the left are phase separation contour phase diagrams and those on the right are the corresponding reproducibility diagrams. Typical values for experiments with and without the incorporation of salt are presented.

Salt Effect Investigations

The effect of salt on coacervate formation was investigated using a layering technique, which is employed with the liquid handler by default due to its dispensing program. Embracing this layering technique, we are able to investigate the interactions of materials with one another before complete sample mixing and the subsequent effect that this has on coacervation at all compositions. In this technique, the materials were added sequentially to the 96-well plate vials, so that Material 1 and Material 2 were allowed to mix via diffusion before Material 3 was added. Materials 2 and 3 were also allowed to mix via diffusion before the entire plate was vortexed to provide complete sample mixing. This pre-mixing of individual ingredients in the 96-well plates provides insight into the effects of introducing the salt first to the surfactant, first to the polymer, and after the polymer and surfactant have interacted. The six possible addition orders for this three component system are listed in Table II. The systems will be discussed in reference to their addition order. For example, Experiment 1 will be referred to as the Salt, Surfactant, Polymer phase diagram.

To investigate salt effects on coacervation two different salt addition methods were employed using the high-throughput screening formulation technique. For the synthetic polymer systems a constant salt method was employed where all points investigated contained the same total amount of salt (30uL). By varying the percent actives of the salt pre-mix we are able to investigate a range of salt concentrations and their effect on coacervation throughout a wide composition range. The studies presented in this research utilize a 5.0% (w/w) pre-mix of salt, which provides 0.05% (w/v) actives at all points in the phase diagram.

Method for Cellulosic Polymers

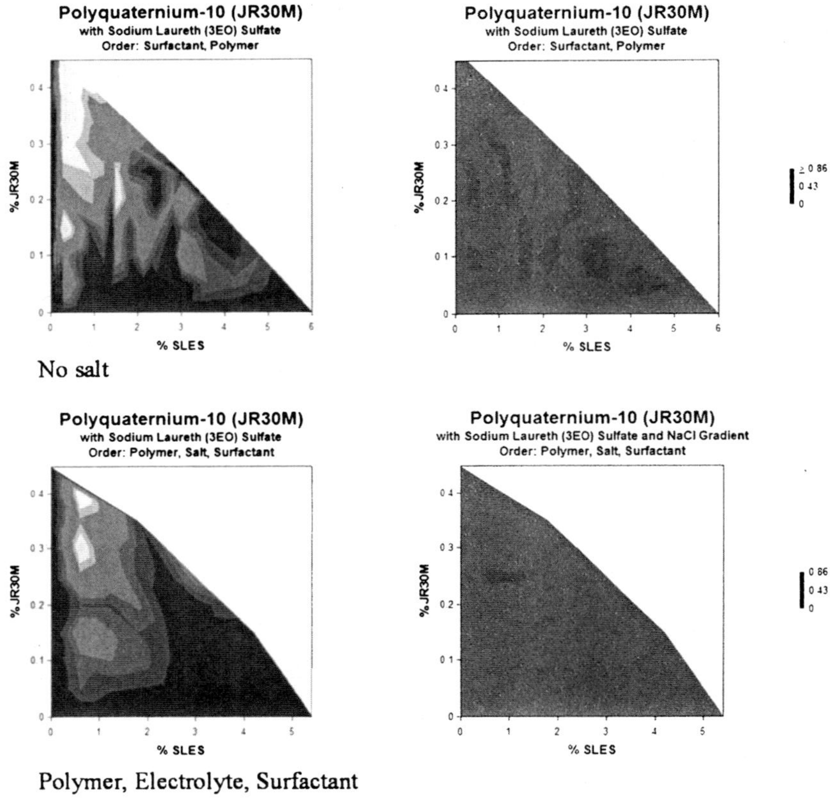

Figure 3. Reproducibility of high-throughput screening methods for cellulosic polymer-surfactant interaction contour phase diagrams.
(See page 4 of color inserts.)

Table II. Addition orders studied.

Experiment Designation	Material 1	Material 2	Material 3
Experiment 1	Salt	Surfactant	Polymer
Experiment 2	Polymer	Surfactant	Salt
Experiment 3	Surfactant	Salt	Polymer
Experiment 4	Surfactant	Polymer	Salt
Experiment 5	Polymer	Salt	Surfactant
Experiment 6	Salt	Polymer	Surfactant

For the cellulosic systems, a salt gradient method was employed where the percent actives in the final sample varies throughout the phase diagram. This method was used for these systems because of its relevance in formulations, where the amount of salt present in a sample would vary depending on the dilution ratio. Six different addition orders were investigated for the cellulosic systems (discussed above). The salt profiles vary slightly between the addition orders, however the variations are so small that the results of all six experiments can by compared. The representative salt profiles are shown in Figure 4.

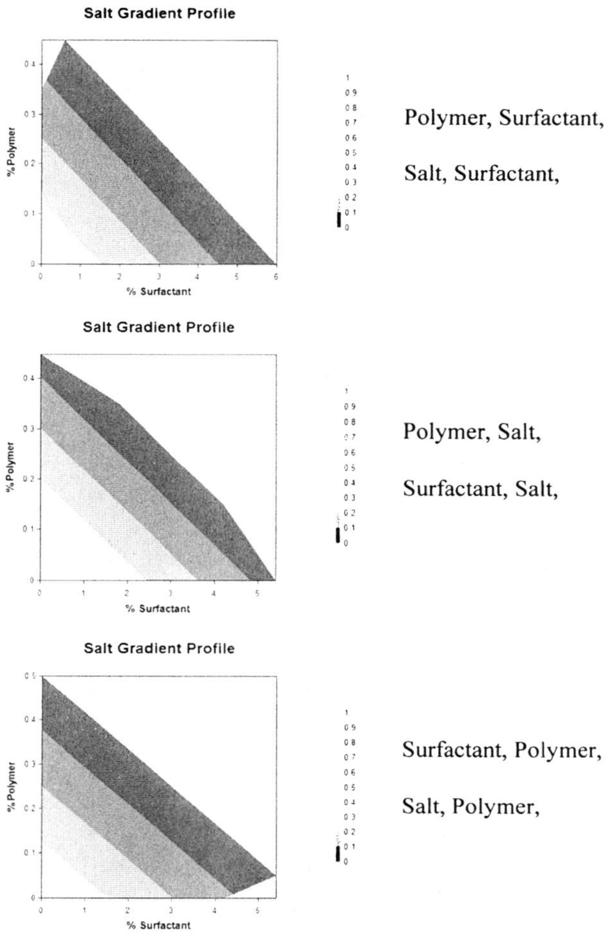

Figure 4. Salt gradient profiles for cellulosic polymer-SLES-NaCl investigations. The addition orders associated with each salt profile are listed to the right of the salt gradient profile phase diagrams.
(See page 5 of color inserts.)

The color gradient in Figure 4 denotes the percent actives (w/v) of NaCl at each composition studied. The salt concentrations range from ~0.5% to very low concentrations, near 0.0% actives.

Results and Discussion

Polymer Structural Effects

Cationic Synthetic Polymers

The impact of polymer structural differences on coacervate formation was investigated using the cationic synthetic polymers poly (4-vinyl pyridine) (P4VP) and poly (2-vinyl pyridine) (P2VP). The structures of these polymers are shown in Figure 5.

Poly (4-Vinyl Pyridine) Poly (2-Vinyl Pyridine)

Figure 5. Cationic synthetic polymer structures.

The contour phase diagrams for the interaction of each of the cationic synthetic polymers with sodium dodecylbenzene sulfonate (LAS) are shown in Figure 6. The data in Figure 6 shows a clear dependence of coacervate amount on polymer structure. Since coacervation is governed by an ion-exchange process, the availability of the cationic group to the anionic surfactant is a key factor in the amount of coacervate produced. The cationic group of P2VP, at the 2-position, is sterically hindered by the polymer backbone which prevents the anionic surfactant molecules from accessing the cationic charge. This is evidenced by the lack of phase separation seen in the P2VP-LAS contour phase diagram (6b). In contrast, the 4-positioning of the cationic group in P4VP provides a highly accessible group for ion-exchange with the anionic surfactant head group. This results in a very high amount of coacervate formation, as seen in the P4VP-LAS phase diagram (6a).

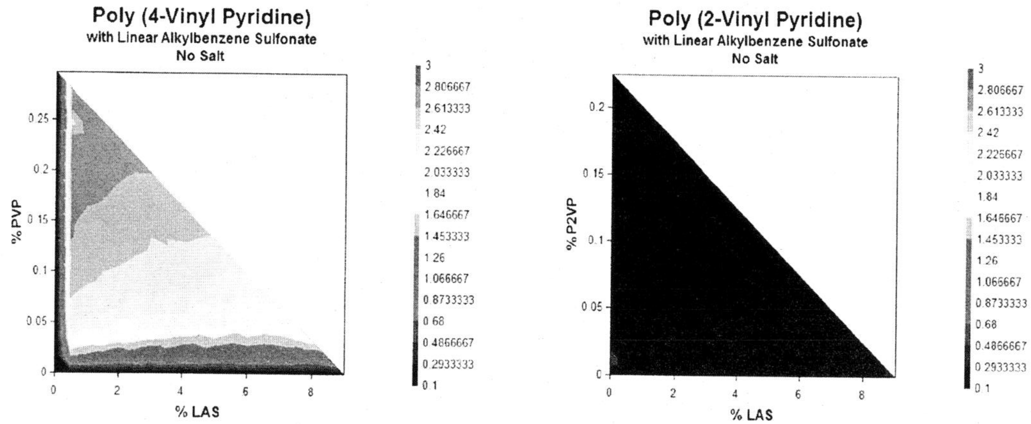

(a) (b)

Figure 6. Contour phase diagrams of (a) P4VP and (b) P2VP with LAS. (See page 6 of color inserts.)

As will be discussed in more detail with the cellulosic polymers, molecular weight is known to have a positive effect on the amount of coacervate formed, where an increase in molecular weight leads to an increase in coacervation (*19,20*). In the P4VP and P2VP systems, the molecular weights of the polymers are 6,717 and 25,762, respectively. Based on molecular weight, there should be increased phase separation with P2VP because it is a much larger polymer. However, in this case, the structural aspects of the polymer override this molecular weight factor.

Cationic Cellulosic Polymers

The effects of molecular weight and charge density on coacervate formation have been previously investigated; however these efforts focus on the dilute regime (*20,21*). Here, investigations of the effects of molecular weight and charge density on coacervation in the semi-dilute and concentrated surfactant regimes were performed. A specific class of cationic cellulosic polymers, polyquaternium-10, was studied; varying in molecular weight and/or charge density, with a constant backbone structure (Table I). The contour phase diagrams for each of the polymers in Table I with sodium lauryl ether (3EO) sulfate (SLES) are shown in Figure 7.

From the data in Figure 7, trends in molecular weight and charge density are readily apparent. The low molecular weight polymers all produce low amounts of coacervate regardless of charge density. At a slightly higher molecular weight and a high charge density, coacervate formation is increased as evidenced by the comparison of the JR125 diagram to the JR400 diagram. At high molecular weights, LR30M and JR30M, coacervate amount is higher than in the low molecular weight systems with the same charge density, LR400 and JR125, respectively. In addition to molecular weight effects at higher molecular weights, an increase in charge density also increases the amount of coacervate produced. The highest amount of coacervate produced is with JR30M, which has the highest molecular weight and the highest charge density. This can be explained using the knowledge of the mechanisms for coacervation in dilute systems: ion-exchange followed by hydrophobic association. In the case of JR30M, the large polymer molecules are occupying more space than the small molecules of the low molecular weight polymers. Thus, when surfactant molecules are bound to a polymer chain via ion-exchange, their hydrophobic tail groups are able to easily associate with bound surfactant tail groups along the same polymer and with those from surfactant bound to other polymer molecules in solution. This can create a network structure which leads to an increased amount of coacervate. In addition, the high charge density will lead to an increased number of bound surfactant molecules which will also contribute to an increased ability of hydrophobic tail association and therefore increased coacervate amount.

Electrolyte Effects

It was of interest for both fundamental understanding and practical use to examine the effects of additional electrolytes on polyelectrolyte-surfactant interactions. Initial studies investigated the effects of NaCl on cationic cellulosic polymer-SLES interactions. For sample preparation via high-throughput screening, the salt was pre-mixed with the surfactant at a 30:1 SLES:NaCl ratio. A representative phase diagram is shown in Figure 8.

In these systems, we saw an overall decrease in the amount of coacervate with the addition of NaCl, as well as a shift in the compositions that produced coacervate. This is likely due to ion-exchange shielding and/or micelle growth. The sodium chloride ions will reside closely to the anionic head groups of the surfactant molecules, causing the anionic charges to be shielded. This would result in a reduced attraction of the oppositely charged polymer and surfactant molecules. In the contour phase diagram for the system containing salt, the lack of separation at low surfactant concentrations, independent of polymer amount, supports the ion-exchange shielding effect of the NaCl. The Na^+ ions that are residing near the anionic surfactant head groups may also be causing shielding of the head group repulsions. This results in a decreased curvature of the surfactant micelles and overall micelle growth. The micelles may grow so large that they dwarf the polymer molecules and coacervation is unable to occur. This could account for the disappearance of phase separation at high surfactant concentrations with the addition of NaCl.

Order of Addition Effects

The initial polymer-surfactant-salt interaction studies prompted a question of the effect of addition order on the coacervation mechanism. If the salt is added first into a polymer environment will the coacervation phenomenon be the same as a system where surfactant and electrolyte are first introduced? This question was addressed by investigating the effect of addition order in the presence of salt for both the synthetic and cellulosic polymer systems. The methods used for each system are discussed in the Experimental section and the addition orders are outlined in Table II.

Synthetic Polymers

The Surfactant, Salt, Polymer and Polymer, Salt, Surfactant experiments were performed on poly (4-vinyl pyridine) (P4VP) to investigate the mechanistic effects of order of addition. The contour phase diagrams are shown in Figure 9. The No Salt phase diagram in Figure 6a will be used as the baseline diagram for understanding the mechanisms associated with salt addition and various addition

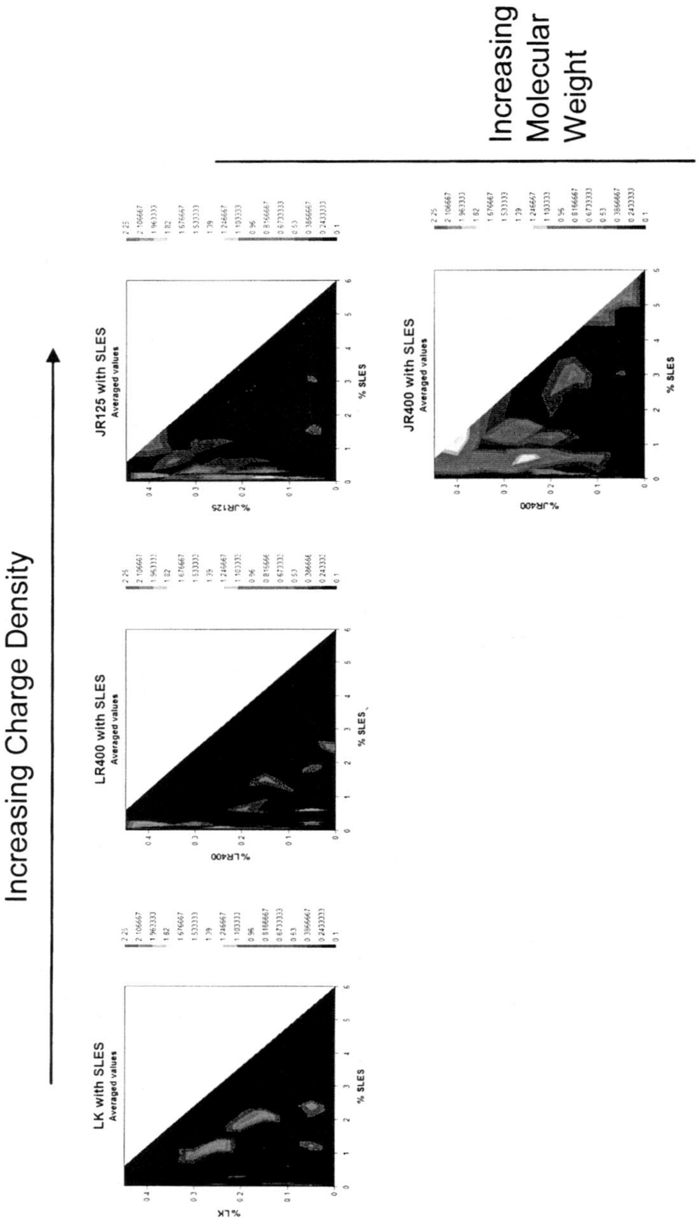
Increasing Charge Density
Increasing Molecular Weight
LK with SLES
Averaged values
%LK
% SLES
LR400 with SLES
Averaged values
%LR400
% SLES
JR125 with SLES
Averaged values
%JR125
% SLES
JR400 with SLES
Averaged values
%JR400
% SLES

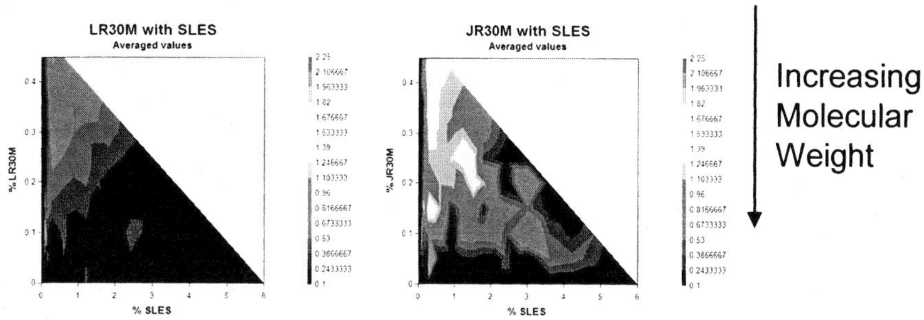

Figure 7. Contour phase diagrams of polyquaternium-10 polymers with SLES. (See page 7 of color inserts.)

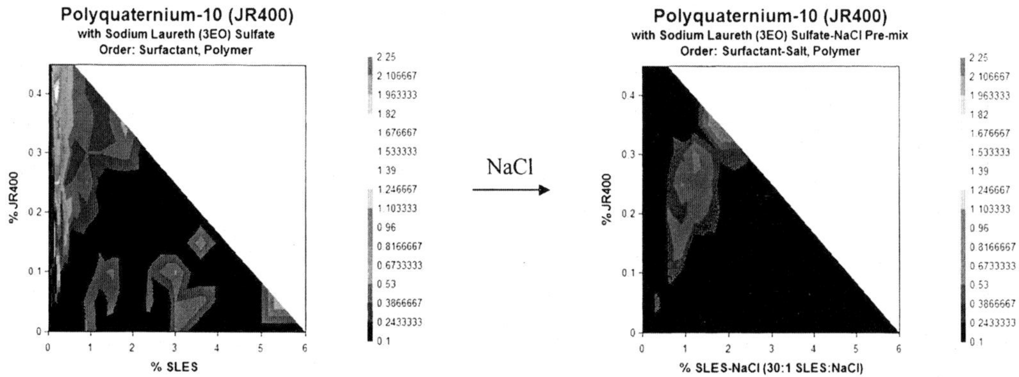

Figure 8. Effect of NaCl on phase separation (NaCl premixed with SLES).
(See page 8 of color inserts.)

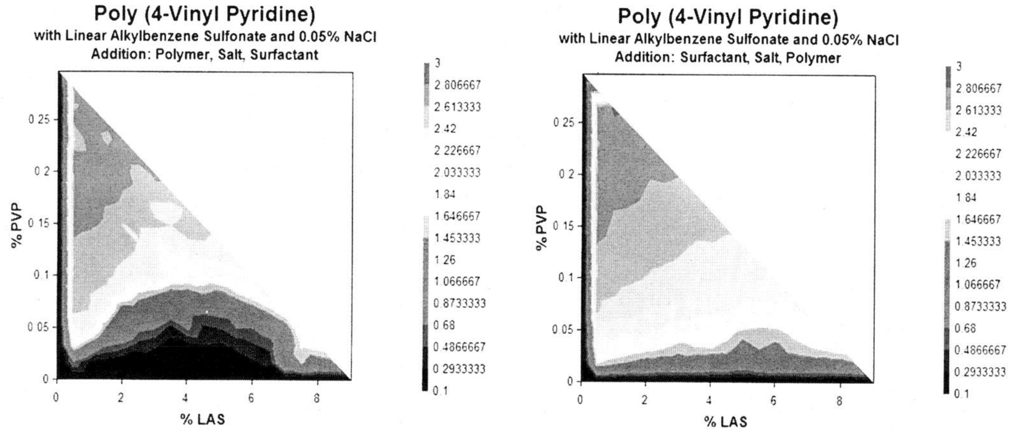

(a) (b)

Figure 9. Phase diagrams for P4VP with LAS (a) Polymer, Salt, Surfactant and (b) Surfactant, Salt, Polymer. (See page 8 of color inserts.)

orders for the synthetic polymers. As previously discussed, the mechanisms of coacervation in the No Salt system are ion-exchange and hydrophobic association. The phase diagrams that do contain salt (9a and 9b) do not contain a salt gradient; there is a constant salt amount for all points on the diagram. In this instance, the salt concentration is 0.05% throughout 9a and 9b.

In the Polymer, Salt, Surfactant phase diagram (9a) polymer collapse is the predominant mechanism. The polymer is mixed with salt before the surfactant is added. This can cause shielding of the cationic groups on the expanded polymer chain from one another, leading to a collapse of the chain. When the chains collapse they become localized areas of extremely high ionic strength (*"super-salts"*) which may cause shielding of the mutual ionic repulsion between surfactant head groups in the micelles. Such a mechanism would lead to pronounced growth in polymer-bound micelle size. This explains the high coacervate amount at low surfactant/low polymer concentrations. However, when these polymers collapse, there is a loss of coacervation at intermediate surfactant concentrations. This is most likely caused by resolubilization, where, in the case of a collapsed polymer, the surfactant is bound by ion-exchange to a more curved region which could sterically shield hydrophobic self-assembled tail associations among bound surfactant molecules. This leads to bound surfactant tail-free surfactant tail hydrophobic interactions at a lower overall surfactant concentration. This is in contrast to the conventional coacervation mechanism where the polymer chain is extended and polymer-bound hydrophobic tail associations increase the amount of coacervation before resolubilization occurs. At high surfactant concentrations micelle growth is the predominant mechanism, with an increase in coacervation in this composition region due to polymeric bridging of large micelles.

The Surfactant, Salt, Polymer contour phase diagram (9b) varies only slightly from the No Salt diagram (6a). At low polymer/intermediate surfactant concentrations there is a slight decrease in the amount of coacervation with the addition of salt. If micelle growth were playing a strong role, an increase in coacervation would be expected in this composition region. This indicates that polymer collapse is the predominant mechanism of coacervation, although it is not as strong as in the latter system due to the mixing of surfactant and salt before polymer is added. There may also be some ion-exchange shielding indicated by a small lessening of coacervate amount in the ion-exchange region. However, due to the similarity of the Surfactant, Salt, Polymer phase diagram to the No Salt diagram, shielding is not a predominant factor.

Cellulosic Polymers

Three cationic cellulosic polymer-SLES systems (JR400, LR30M, and JR30M) were examined using all addition orders to investigate whether the mechanistic effects of order of addition seen with the synthetic polymer systems are universal or system specific. The trends of molecular weight and charge density that were presented in Figure 7 also existed between these polymer systems with salt. The data from JR30M with SLES will be presented to represent these trends because it has the overall highest amount of coacervation. For reference, the JR30M-SLES contour phase diagram with no salt added is shown in Figure 10.

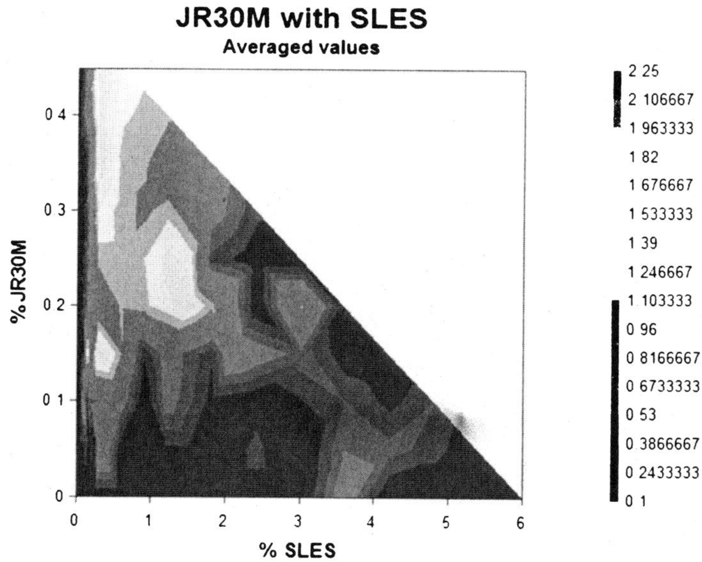

Figure 10. Contour phase diagram of JR30M-SLES with no salt. (See page 7 of color inserts.)

From previous research, it is know that the coacervation mechanism for the system in Figure 10 is ion-exchange and hydrophobic association. Given this information, the No Salt diagram can be used as a baseline for understanding the effects of salt order of addition.

Ion-Exchange and Micelle Growth

The addition orders of Salt, Surfactant, Polymer (Experiment 1) and Polymer, Surfactant, Salt (Experiment 2) exhibited ion-exchange and/or micelle growth coacervation mechanisms. The contour phase diagrams for both addition orders are shown in Figure 11. The addition orders presented in Figure 11 include salt gradients as discussed in the Experimental section. The salt gradient for both Figures 11a and 11b is shown in Figure 4.

Comparison of the contour phase diagrams 11a and 11b shows that there is a definite effect of salt addition order on coacervate formation. In the Salt, Surfactant, Polymer experiment (11a) ion-exchange and micelle growth are the predominant mechanisms. High amounts of coacervate are seen at low surfactant concentrations, similar to the No Salt phase diagram (Figure 10), however the range of high coacervation is shifted to slightly higher surfactant concentrations in the experiment with salt. This indicates a slight shielding effect of the electrostatic interactions by the salt, where a greater amount of surfactant is required to achieve the same amount of coacervate. Micelle growth is also possible since the salt and surfactant were allowed to mix before polymer is added. Comparing with the No Salt diagram, micelle growth is evidenced by the increase in coacervation in regions of intermediate surfactant concentration.

The Polymer, Surfactant, Salt phase diagram (11b) shows overall less coacervate than the Salt, Surfactant, Polymer diagram (11a) and the No Salt diagram. In this experiment, the polymer and surfactant were allowed to mix before salt was added, which promotes ion-exchange interactions. However, once the salt was added, the chemical potential of the system shifts such that ion-exchange is reversed which leads to a resolubilization of the coacervate formed via electrostatic interactions at some compositions.

Micelle Growth and Micelle Bridging

The addition order of Surfactant, Salt, Polymer (Experiment 3) exhibited micelle growth and micelle bridging coacervation mechanisms. The contour phase diagram is shown in Figure 12. The addition order presented in Figure 12 includes a salt gradient as discussed in the Experimental section. The salt gradient for Figure 12 is shown in Figure 4.

In the Surfactant, Salt, Polymer experiment micelle growth and micelle bridging are equally strong as coacervation mechanisms. Electrostatic interaction is present, as evidenced by the high coacervate amounts in the low surfactant/high polymer region, however compared with the No Salt diagram (Figure 10), these interactions are restricted due to shielding by the salt added as the second material, so that polymer and surfactant to not directly interact until complete sample mixing. Micelle growth occurs in the surfactant/salt layer, leading to increased coacervation at intermediate surfactant concentrations. In addition to larger micelles, there is a possibility of bridging between micelles by the polymer chains. This class of cellulosic polymers has a stiff backbone with pendant cationic groups, therefore, it is unlikely that the addition of NaCl would

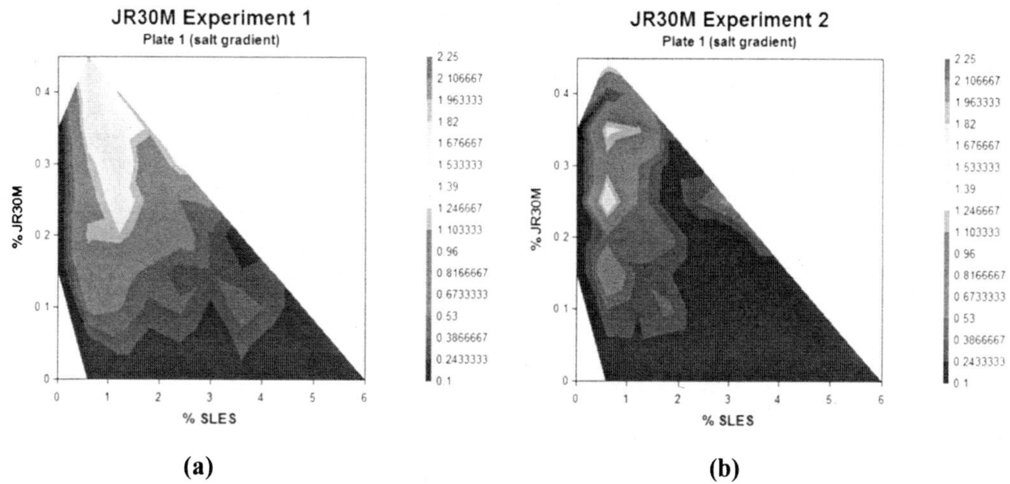

Figure 11. Contour phase diagrams for JR30M-SLES-NaCl with (a) Salt, Surfactant, Polymer and (b) Polymer, Surfactant, Salt addition orders.
(See page 8 of color inserts.)

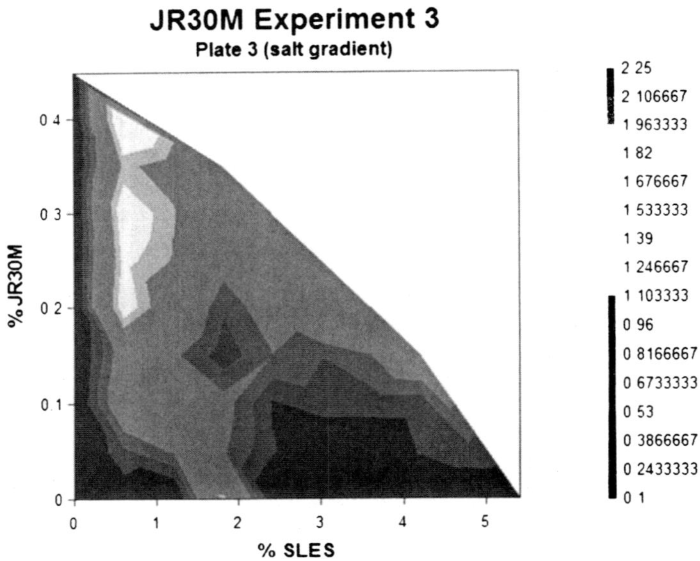

Figure 12. Contour phase diagram for JR30M-SLES-NaCl with Surfactant, Salt, Polymer addition order. *(See page 8 of color inserts.)*

cause complete collapse of the polymer. However it is possible that the salt would shield repulsions between the cationic groups along one chain and among neighboring chains. This shielding could lead to cationic sites in closer proximity to one another, which would in turn promote hydrophobic associations among surfactant tail groups bound to the same polymer chain and possibly neighboring polymer chains. These intra- and inter-micellar bridges could lead to large amounts on coacervate being produced in regions that showed little to no coacervate in the absence of salt.

Micelle Bridging

The addition order of Polymer, Salt, Surfactant (Experiment 5) exhibited only a micelle bridging coacervation mechanism. The contour phase diagram is shown in Figure 13. The addition order presented in Figure 13 includes a salt gradient as discussed in the Experimental section. The salt gradient for Figure 13 is shown in Figure 4.

In the Polymer, Salt, Surfactant experiment, micelle bridging is the predominant mechanism. Shielding of the cationic charge repulsions occurs in the polymer/salt layer before surfactant is added. Upon addition of surfactant, hemi-micelles and micelles form along the polymer chain and inter- and intra-micellar tail group associations can occur, leading to a coacervate profile similar to that in Figure 12 at low surfactant compositions. However, micelle growth is not predominant in the Polymer, Salt, Surfactant experiment, leading to reduced coacervation at intermediate surfactant concentrations compared to Figure 12. Also, the addition of salt as the second material increases the predominance of electrostatic interaction shielding so ion-exchange is not predominant in this experiment.

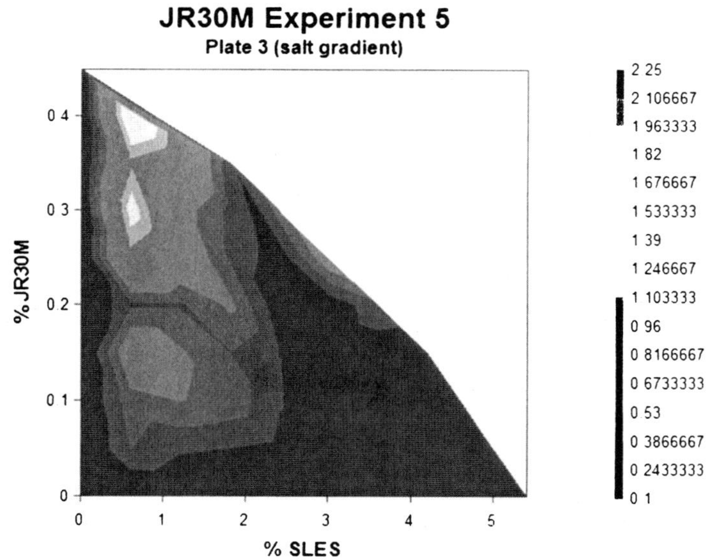

Figure 13. Contour phase diagram for JR30M-SLES-NaCl with Polymer, Salt, Surfactant addition order. (See page 9 of color inserts.)

Ion-Exchange and Micelle Bridging

The addition orders of Surfactant, Polymer, Salt (Experiment 4) and Salt, Polymer, Surfactant (Experiment 6) exhibited ion-exchange and/or micelle bridging coacervation mechanisms. The contour phase diagrams are shown in Figure 14. The addition orders presented in Figure 14 include salt gradients as discussed in the Experimental section. The salt gradient for both Figures 14a and 14b is shown in Figure 4.

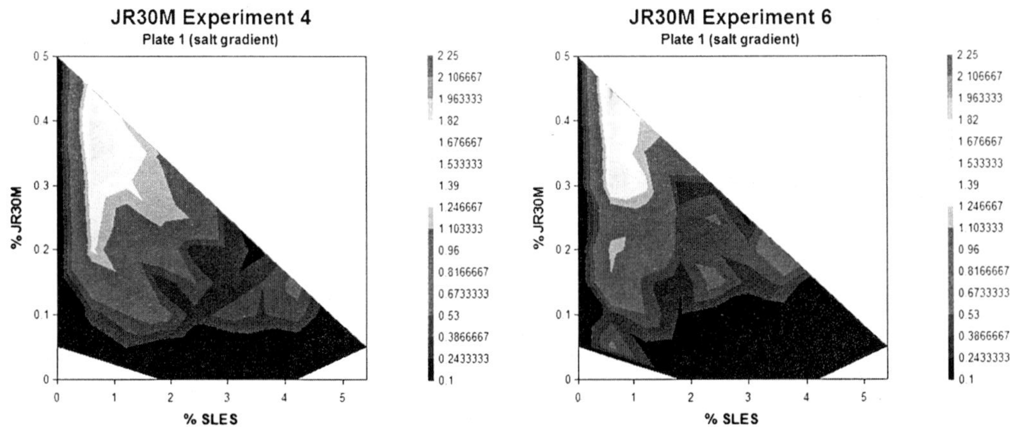

(a) **(b)**

Figure 14. Contour phase diagrams for JR30M-SLES-NaCl with (a) Surfactant, Polymer, Salt and (b) Salt, Polymer, Surfactant addition orders.

(See page 9 of color inserts.)

In the Surfactant, Polymer, Salt (14a) experiment, ion-exchange and micelle bridging are both predominant mechanisms. A surfactant/polymer layer is formed before salt is added leading to a large amount of coacervate due to ion-exchange, as seen in the low surfactant/high polymer region of the diagram. Although the salt is added to a system where electrostatic interactions have already occurred, resolubilization is not observed as it was in Figure 11b. This indicates that the influence of salt in this system is likely shielding of intra- and inter-molecular cationic repulsions, which promotes micelle-bridging and leads to increased coacervation at intermediate surfactant concentrations.

In the Salt, Polymer, Surfactant (14b) system the polymer and salt are mixed before the surfactant is introduced. This allows for micelle-bridging to become a stronger mechanism than ion-exchange, giving a phase diagram with regions of coacervation, independent of amount, similar to that in Figure 13. In contrast to Figure 13, where only micelle bridging occurred, the Salt, Polymer, Surfactant system also exhibits an ion-exchange mechanism due to the addition of salt as the first material. This reduces the amount of electrostatic interaction shielding as evidenced by the high coacervate amount in the low surfactant/high polymer region of Figure 14b.

Conclusions

The high-throughput screening formulation method that had previously been developed in our research group has allowed the understanding of structure-property relationships and coacervation mechanisms in the semi-dilute and concentrated surfactant regimes.

Using synthetic polymers we have determined that the positioning of the cationic group along the backbone chain can impact the amount of coacervate formed. The more available the cationic group is to the surfactant molecules, the greater the amount of coacervate produced. In addition to structural effects, we have determined the impact of molecular weight and charge density on coacervation within the cationic cellulosic polyquaternium-10 polymer series. Using contour phase diagrams, we observed that an increase in molecular weight leads to an increased amount of coacervation. Similarly, an increase in charge density leads to an increased amount of coacervation.

Lastly, we have determined that in addition to polymer structure, the addition order of materials, such as polymer, surfactant and salt, to a formulation affects the compositional range of coacervate formation and the coacervation mechanism. Depending on the flexibility of the polymer structure, the mechanisms of coacervation in the presence of added electrolyte can vary. Poly (4-vinyl pyridine)-LAS-NaCl systems show a polymer collapse mechanism with the formation of a *"super-salt"* when salt and polymer are pre-mixed. However, for the stiff cellulosic polymers, the pre-mixing of salt and polymer before surfactant addition enhances micelle-bridging without complete polymer collapse. Also, the addition of salt to surfactant with the cellulosic systems induces micelle growth and subsequent enhanced coacervation at intermediate and high surfactant concentrations due to bridging of the surfactant by polymer molecules.

This knowledge of polymer structure-coacervate property relationships, as well as the understanding of addition order effects can guide the formulator to better products and a better understanding of the interactions of the formulation materials.

References

1. Goddard, E. D., Ananthapadmanabhan, K. P., eds. *Interactions of Surfactants with Polymers and Proteins*; CRC Press: Boca Raton, FL, 1993.
2. Hayakawa, K. K., J. C. T. *Journal of Physical Chemistry* **1983**, *87*, 506-509.
3. Leung, P. S. G., E. D.; Han, C.; Glinka, C. J. *Colloids and Surfaces* **1985**, *13*, 47-62.
4. Wang, C. T., K. C. *Langmuir* **2002**, *18*, 6484-6490.
5. Yamato, N. K., R.; Lochhead, R. Y. In *Proceedings of the International Conference on Colloid and Surface Science*: Tokyo, Japan, 2000.
6. Bakshi, M. S. S., S. *Colloids and Polymer Science* **2004**, *282*, 993-999.
7. Bastardo, L. A. G., V. M.; Bergstrom, M.; Claesson, P. M. *Journal of Physical Chemistry B* **2005**, *109*, 167-174.
8. Chang, Y. M., C. L. *Polymer Preprints* **1993**, *34*, 992-993.
9. Chronakis, I. S. A., P. *Macromolecules* **2001**, *34*, 5005-5018.
10. Guillot, S. D., M.; Desert, S.; Langevin, D. *Langmuir* **2003**, *19*, 230-237.
11. Jain, N. T., S.; Guillot, S.; McLoughlin, D.; Langevin, D.; Letellier, P.; Turmine, M. *Langmuir* **2004**, *20*, 8496-8503.
12. Kogej, K. T., E.; Reynaers, H. *Langmuir* **2002**, *18*, 8799-8805.
13. Regismond, S. T. A. H., Y.; Goddard, E. D.; Winnik, F. M. *Langmuir* **1999**, *15*, 3007-3010.
14. Vincekovic, M. B., M.; Smit, I,; Filipovic-Vincekovic, N. *Colloids and Surfaces, A* **2005**, *255*, 181-191.
15. Wang, X. L., Y.; Li, J.; Wang, J.; Wang, Y.; Guo, Z.; Yan, H. *Journal of Physical Chemistry B* **2005**, *109*, 10807-10812.
16. Svensson, A., Sjostrom, J, Scheel, T., Picullel, L. *Colloids and Surfaces, A* **2003**, *228*, 91-106.
17. Convertine, A. J., Sumerlin, B. S., Thomas, D. B., Lowe, A. B., McCormick, C. L. *Macromolecules* **2003**, *36*, 4679-4681.
18. *Cationic conditioners that revitalize hair and skin at www.dow.com/PublishedLiterature* **2002**.
19. Li, Y. X., J.; Dubin, P. L. *Macromolecules* **1994**, *27*, 7049-7055.
20. Wang, Y. K., K.; Dubin, P. L. *Macromolecules* **2000**, *33*, 3324-3331.
21. Hansson, P. A., M. *Journal of Physical Chemistry* **1996**, *100*, 9038-9046.

Advances in Eye Care

Chapter 6

Beauty from Contact Lenses beyond Vision Correction

Michael H. Quinn and Walt Bassett

CIBA Vision, 11460 John's Creek Parkway, Duluth, GA 30097

Contact lenses have recently moved from passively enhancing beauty by their role in replacing spectacles to becoming active beauty aids. The active beauty aid modality occurs when contact lenses change or enhance the color of the wearer's eyes. Eye color changes with contact lenses can be dramatic when a person with dark eyes appears to have blue eyes. Dramatic eye color changes occur by placing an appropriate, intermittent pattern of opaque pigments on the contact lens that overlays the iris. Underlying iris color aids the pigment in changing eye color. In addition, subtle enhancements are available whereby a person with blue eyes appears to have green eyes. Reactive dyes can make cosmetic lenses that enhance rather that change eye color. Chemistry and technology behind these apparent eye changes will be discussed.

Introduction

The first commercially successful contact lenses were made from PMMA (polymethyl methacrylate). Their success was limited because the wearer's eye shape had to adapt to the shape of the lenses. These lenses were useful for correcting vision but were uncomfortable due not only to their rigidity but also to their minimal ability to transport the amount of oxygen needed to maintain eye health (*1*). Polymethyl methacrylate lenses are still worn but modern rigid contact lenses are generally made from silicone based materials capable of transporting significantly more oxygen. This increases the probability that the eye will receive sufficient oxygen to maintain health. However, rigid lenses are still generally uncomfortable.

Many rigid contact lenses contain a lightly colored, edge-to-edge visibility tint to aid in inserting them as well as finding them on a sink or carpet. Their relatively small diameters (about 8-10 mm) do not allow them to make significant changes to eye color since this is barely the size of the iris that can reach up to 10 mm depending on the person and illumination. Furthermore, all contact lenses are designed to move on the eye. Lens movement hinders their ability to change eye color since the edges of the iris would be inconsistently covered by the colored portions. Colorant is usually added to rigid lenses by entrapping dyes or pigments during the polymerization process. There is little chance that either dye or pigment colorants will be eluted out of the rigid lens polymer matrix while they are worn since tears do not penetrate the lenses.

Most current commercially available contact lenses are made from hydrophilic materials (hydrogels) that permanently retain water when they are worn. These lenses are flexible enough to accommodate many differences in eye contour among individuals. They are more comfortable than rigid lenses and this has accounted for their wide acceptance. Hydrophilic clear contact lenses have been commercially available for several decades.

The first hydrophilic lenses were made primarily from HEMA (2-hydroxyethyl methacrylate) and contained about 38 % water (*2*). The lenses also contained a small amount of a crosslinking monomer. Water content was increased in later formulations by adding other monomers such as methacrylic acid. Hydrophobic monomers were often added to provide strength. Other hydrophilic materials such as N-vinyl pyrrolidone, polyvinyl alcohol and glyceryl methacrylate have also been used to make hydrophilic contact lenses.

Oxygen is transported through water in hydrogel lenses (*3*). The amount of oxygen transported is proportional to the amount of water in the lens. Increased oxygen transport is associated with increased health benefits. However, the water content of lenses from these materials is limited and hence the amount of oxygen transported is limited. Unfortunately, the amount of transported oxygen is less than that desired.

Oxygen transport has been increased in newer materials that combine hydrophilic monomers with silicone and perhaps fluorine containing monomers

(*3*). These lenses have both hydrophobic and hydrophilic characteristics since they still contain water. The primary oxygen transport medium is through the silicone containing parts of the lens and water is a secondary medium.

Hydrophilic contact lenses offer a suitable platform to alter eye color since they have diameters between about 13.5 – 14.5 mm which allow them to cover most or all of the iris when the lens moves during wear. A hurdle to make colored hydrophilic contact lenses has been adding colorant that will not elute from the lenses. This has been solved either by adhering a printed pattern onto the contact lens or by reacting dyes with the contact lens base polymer. The printed pattern is generally composed of opaque pigments while reactive dyes are translucent.

Colored Contact Lenses

Hydrophilic contact lenses containing colorant fall into the categories of (1) handling tints and (2) cosmetic colored lenses. Cosmetic colored lenses are divided into tints to enhance light eyes and opaque prints to change eye color.

Handling Tints

Handling tints for hydrophilic lenses are analogous to the lightly colored edge-to-edge visibility tints in PMMA and successive generations of rigid lenses. They are usually made by dispersing pigment in the monomer mixture before polymerization to form contact lenses. Pigment is entrapped in the resulting polymer network. Pigments are generally used rather than dyes to prevent elution during wear. Pigment particle size is reduced sufficiently prior to its addition to the monomer mixture so as not to interact with light thereby causing optical aberrations. A patient's visual acuity while wearing a tinted lens is comparable to a clear lens. Tint aids in inserting lenses and finding dropped lenses. Tint level is low enough that lens color on the eye is not apparent to a casual observer.

Cosmetic Colored Lenses

Tints to Enhance Light Eyes

Enhancing contact lenses are used to give the person wearing them a subtle enhancement to their eye color. For example, a blue eye could become another shade of blue or become green or aqua. These lenses usually do not give acceptable cosmetic effects to people with darker eye colors.

Early work to make hydrophilic contact lenses with enhancing tints was done by exposing them to aqueous dye solutions (*4*, *5*). Dyes were not permanently bound and eluted from the lenses. This was unacceptable and development work focused on other approaches.

The technology quickly evolved to methods that permanently bonded dye to contact lenses. In one approach, HEMA was polymerized with a monomer such as vinylethyl maleate that can be photochemically reacted with diazonium double salts (*6*). The latter reaction permanently bonded dyes to the contact lens.

Pendant hydroxyl groups of HEMA units in pre-formed poly-HEMA contact lens polymers have been used to make tinted contact lenses (*7*). An aqueous solution of a reactive sulfonyl dyestuff was imbibed into a contact lens and the binding reaction completed by exposing it to sodium carbonate. Dye can be imbibed into the lens in a pattern. For example, the peripheral edge of the lens could remain clear while the central part covering the iris could be tinted. This resultant lens is tinted over the iris but not over the white sclera. In a variation of this process, the same reactive dyestuff can be reacted to monomeric HEMA prior to its use to make a contact lens (*8*). An example of the dyes used in either method is shown in Figure 1.

Enhancing tints can be used for more than cosmetic effects and have a practical purpose. An example is the ProSoft contact lens. The colorant on the lens is a combination of various dyes that allow the lens to preferentially transmit specific wavelengths corresponding to a desired color, Figure 2. In this example, the transmitted color is similar to that of a tennis ball and other wavelengths are diminished. This should allow the tennis ball to standout from the background. The perceived higher contrast between the tennis ball and everything else should allow the tennis player to more readily focus on the tennis ball.

Opaque Prints to Change Eye Color

Otto Wichterle, inventor of hydrogel contact lenses, first disclosed (*9*) the concept of going beyond enhancing eye color to actually changing apparent eye

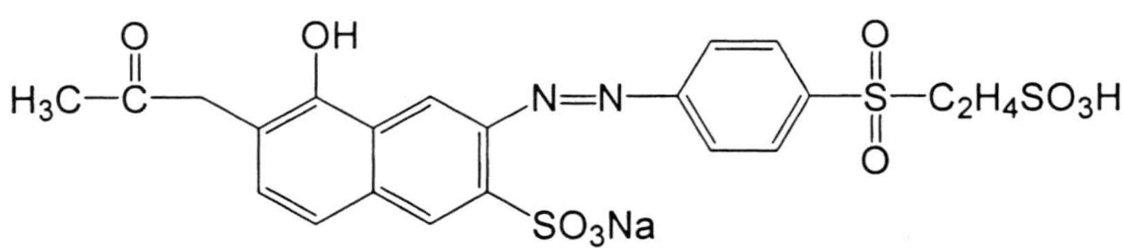

Figure 1. Example of a reactive dye used to make enhancing contact lenses.

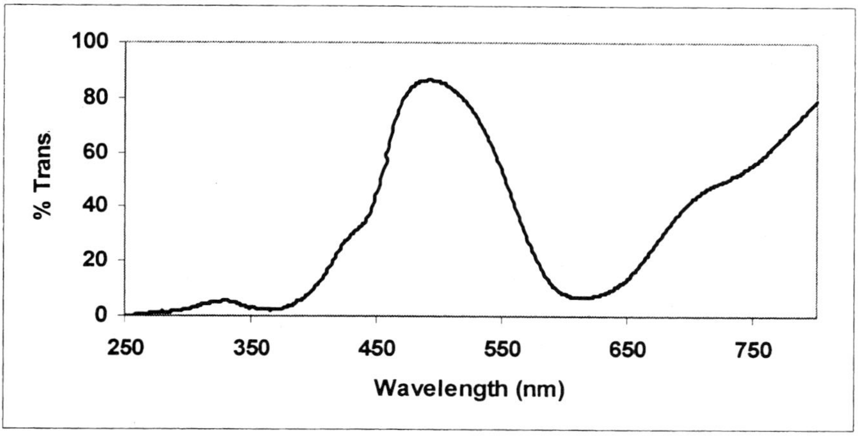

Figure 2. Spectrum of ProSoft contact lens showing preference to transmit "tennis ball green" light.

color by making a pigmented rather than tinted contact lens. His method involved drying a hydrated lens then flattening it on a plate prior to adding the pigment in a pattern resembling the iris. This was not a practical method and did not achieve commercial success.

A pigmented eye color changing hydrogel lens was introduced commercially (10). This lens was made by polymerizing a thin, pigmented layer into clear polymer rod. This was lathe cut resulting in a lens with a thin pigmented layer between two relatively thick clear layers. Lens curvature allowed a clear optical zone to be formed because the pigmented layer was relatively thin. The lens did change eye color but was not a commercial success because the pigmented layer was solid and did not resemble a real iris.

The first commercially successful pigmented, opaque print on a contact lens was made by adhering pigment on a lens surface (11). This advance made a contact lens that contained an intermittent, colored, opaque pattern over the iris. The intermittent pigmented pattern left much of the lens covering the iris uncolored allowing the underlying iris to contributed to color change.

Cosmetic lenses have evolved beyond the use of a single color and pattern (*12*). For example, all the patterns in Figure 3 would be individual colors and applied to a single contact lens yielding a product that combines colors and patterns to closely resemble a human iris. These patterns still contain a significant amount of uncolored space amongst the pigmented pattern. Patterns, colors and uncolored spaces have been combined to make a commercially successful product than gives the eye a natural appearance.

Figure 3. Intermittent, pigmented patterns on a single hydrogel contact lens.

The three patterns, Figure 3, make a contact lens that changes eye color while maintaining a natural appearance. However, a natural appearance is not always the effect desired by truly leading edge, fashion conscious consumers. This group of people can go beyond a natural appearance with products such as Wild-Eyes®, Figure 4.

Figure 4. Examples of leading edge fashion contact lenses.

Another type of colored contact lens is used as a prosthetic device by people with damaged eyes. These lenses combine features of conventional colored lenses and technology used to manufacture Wild-Eyes®. These lenses have a conventional, natural appearing print that may be combined with a completely occluded pupil or one that may be otherwise restricted (*13, 14, 15*). These lenses allow those who need them to appear in public without drawing attention to a disfigured eye.

Pigments have been attached to lenses in many ways. For example, they can be entrapped in/ on the lens using a reactive ink (*16*). Ink is formulated with pigments, binder polymer with reactive groups such as hydroxyls, solvents and an adhesion promoter such as a diisocyanate. Pigment particle size is reduced to a suitable size prior to ink formulation. Ink is placed on a contact lens, penetrates it and is cross linked thereby entrapping the pigment in an interpenetrating network. In later versions of this technology, the ink is placed on a mold prior to casting a contact lens in the mold (*17*).

Patterns are placed onto lenses or casting molds by dispensing excess ink onto a plate etched with an appropriate pattern. Excess ink is removed but still retained in etched portions. A soft rubber pad is pressed against the plate to transfer much of the ink in the etched portions to the pad. The pad is then pressed against the contact lens or the lens casting mold. Much of the ink from the pad is transferred to the contact lens or the lens casting mold. Pigment can also be placed on a lens with ink jet printers (*18*).

Printing pigmented inks onto contact lenses or onto casting molds prior to lens formation is the most prominent commercial method to make opaque colored lenses. However, other techniques have been developed to make opaque colored lenses. A method was disclosed whereby an aqueous barium chloride solution was absorbed into a hydrated lens in a selected pattern (*19*). The lens was then exposed to sulfuric acid to form insoluble, opaque barium sulfate in the lens. Color was added by reacting the lens polymer with conventional reactive dye techniques.

Another method to add opaque pigment to a hydrophilic lens is to soak the lens in a solution of a water soluble vat dye (*20*). The dye can be imbibed into the lens in a pattern. After the dye is soaked into the lens, it is treated with sodium nitrite to oxidize sulfate ester groups on the dye. The dye is now insoluble in water and permanently entrapped in the lens polymer matrix.

Pigments Used in Cosmetic Contact Lenses

Pigments make colored parts opaque and uncoated parts allow the underlying iris to contribute to color change. Pigments were chosen because of their relatively inert nature and good ability to render an surface opaque. Furthermore, they have a long history of successful use in the printing, paint and plastics industries.

Inorganic, metalo-organic and organic pigments are key components of opaque cosmetic contact lenses (*16*) and will be discussed in greater detail. Examples of inorganic pigments are titanium dioxide, a variety of iron oxide pigments and pearlescent pigments. Blue and green phthalocyanides are metalo-organic pigments and carbazole violet is an example of an organic pigment.

Titanium dioxide, a colorless, crystalline solid, has long been one of the most widely used pigments in printing. Because it strongly reflects most visible light wavelengths, it appears white to the eye. It is also an excellent opacifying material. Its opacifying strength depends on three properties, 1)refractive index, 2)particle size and 3)overall concentration in the polymer. The first is a fundamental property of the particular crystal size. The second, particle size, is controlled to deliver the appropriate opaquing capability. Pigment loading , or concentration is then adjusted in order to balance the need for a cosmetic effect with any tendency to affect desired, polymer mechanical properties. Optimum crystal size for TiO_2 pigments in plastics are between 170 to 240 nanometers.

The family of iron oxide pigments allows us to bring contrast and color into designs. Naturally occurring oxides provide a variety of choices among reds, yellows, browns and blacks. Particle sizes between 100 nanometers to 6 microns are usually found in common use.

Other inorganics are the pearlescent class of pigments (*21*). Pearlescent pigments appear opalescent because they are sources of multiple light reflection. One example of pearlescent pigments are flat particles of mica encapsulated in titanium dioxide. The multiple reflection of light from such a particle add an impression of depth. These pigments are relatively large and can range from 2 to about 110 microns. Carbazole violet is an organic pigment used in opaque lenses as are the metalo-organic pigments phthalocyanine blue and green.

Critical Pigment Volume Concentration

Physical and mechanical properties of the bulk contact lens are extremely important in providing patient satisfaction and performance. These affect comfort and visual acuity experienced by the patient. These critical factors can be affected by the addition of pigments to the polymer. Pigments, of course, also affect the cosmetic effect. Pigment loading then becomes a key factor for appeal and wearability. A balance is needed between having enough pigment loading to obtain the desired colors and designs but at the same time keeping concentrations low enough to avoid negatively affecting lens performance. Molock (*22*) et al. helps us understand how this applies to contact lens design.

The paint and coatings industry has long understood the interactions of pigments, binders, air, solvents and polymer matrixes. A broad description on the current knowledge in this field, can be found in Patton (*23*) as he details pigment dispersion and the concurrent properties derived by various formulations. Pigment volume concentration (PVC) obviously can vary between 0 and 100%. Therefore all binder and no pigment is a PVC = 0%, and all pigment no binder = 100%.

In creating a mixture of pigment / binder there are three phases to consider, air, pigment and binder. As you add binder to the pigment, the liquid displaces air until a point is reached where the air is totally displaced with binder. This defines the CPVC (Critical Pigment Volume Concentration). Beyond this point, further addition dilutes the pigment and the particles are allowed to disperse in the increasing liquid volume. The idea of CPVC is then best understood in the following way. Below the CPVC pigment particles are dispersed widely through the system, while above the critical volume, they are closely packed and contain air pockets.

Why this value is important in creating pigment-binder system comes from many empirical observations on the basic properties of the chemical system were are trying to color. Basically three types of properties are most important;

- Physical
- Optical
- Permeability

Physical properties such as Refractive Index, Density, Modulus, Strength, Impact resistance etc. are critical for on-eye comfort and dimensional stability Obviously any negative impact on optical quality that effects visual acuity will not be acceptable. However, in creating natural cosmetic designs, hiding power, tinting strength, gloss and light scattering properties need to be factored into the aesthetic designs.

Permeability has to be considered due to the fact that the majority of cosmetic contact lenses are made from hydrogel polymers and there is added factor of tear fluids and lens care solutions included in the overall polymer matrix.

Pigment Volume Concentration

Pigment Volume Concentration (PVC) is simply the volume fraction of pigment versus binder. (without consideration of air, or bulk density). The Pigment Packing (φ) factor takes into consideration of the entire volume. Therefore at CPVC and below, φ = PVC.

In paints a measure of surface finish from matte to high gloss exists. Reduced Pigment Volume Concentration is defined as A = PVC/CPVC. Other key factors optimizing pigments for colors and designs involve;

- particle size
- particle shape
- surface area
- porosity
- aglomeration size / structure
- particle size distribution
- chemical functionality

Each one is a topic beyond the scope of this discussion, but serve as a guide to further investigation in this area. Figure 5 is a visual rendering of the three possible conditions of pigment, binder and air situations.

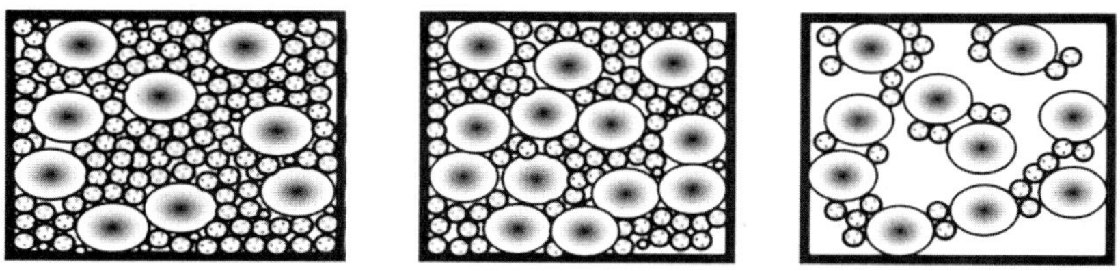

Figure 5. L) PVC<CPVC; M) PVC=CPVC; R) PVC>CPVC; Large circles represent pigment particles and small represent binder.

Safety Considerations

Only FDA approved color additives may be used for coloring contact lenses, and FDA device marketing approval is also needed for colored lenses prior to commercialization in the U.S. Contact lenses, even if used for cosmetic reasons, are medical devices and must be prescribed by a licensed eye care professional. In addition, Federal law (U.S.) prohibits dispensing of contact lenses without a prescription.

The eye care professional will recommend the proper care, wear and replacement regimen of your contact lenses. Improper use of lenses can lead to ocular complications requiring medical treatment. Sharing contact lenses may spread micro-organisms that could result in serious eye health problems.

References

1. Smith, B.J., Fink, B.A. and Hill, M.M., Contact Lens Spectrum, **1999**, Jan., 31.
2. Wichterle, O. and Lim, V., U. S. Patent 2,976,576, 1961.
3. Alvord, L., Court, J., Davis, T., Morgan, C. F., Schindhelm, K., Vogt,. J. and Winterton, L., Optometry and Vision Science, **1998**, vol 75(1), 30.
4. Foley, W. M., Jr., U. S. Patent 4,252,421, 1981.
5. Newcomer, P. C. and Janoff, L. E., Amer. J. of Optometry and Physiological Optics, **1977**, vol 54(3), 160.
6. Tanaka, T, Takahashi, K. and Yoshikawa, T., U. S. Patent 4,157,982, **1987.**
7. Su, K. C., U. S. Patent 4,468,229, 1984.
8. Foley, W. M., Jr, U. S. Patent 4,252,421, 1981.
9. Wichterle, O., U. S. Patent 3,679,504, 1972.
10. Kerr, C. Optician, **1987**, vol 194, 47.
11. Knapp, J. Z., U. S. Patent 4,582,402, 1986.
12. Ocampo, G. J., U. S. Patent 6,827,440 B2, 2004.
13. Anderson, A. E., Spectrum, Dec. **1995**, 33.
14. Spinnell, M. R. and Santilli, J., Spectrum, May **1992**, 17
15. Sinnokrot, A., ICLC, **1996**, 23, 72
16. Loshaek, S., U. S. Patent 4,668,240, 1987.
17. Rawlings, D. L., U. S. Patent 5,116,113, 1992.
18. Tucker, R.C., U.S. Patent 6, 902, 274, *2005*.
19. da Costa, N. M., U. S. Patent 4,981,487, 1991.
20. Jenkins, J. E, WO 87/02079, 1987.
21. Quinn, M. H. and Atkins, B. L., U. S. Patent 6,196,683, 2001

22. Morlock, F.F., Chedhab, K., Ford, J.D., Kindt-Larsen, T., Walker, C.W., Wood, J.M.,Weber, J.E., U.S. Patent Application US2002/0133889 A1, **Sept. 26, 2002**
23. Patton, T.C. In *Paint Flow and Pigment Dispersion, 2nd Ed.*, Wiley-Interscience, **1979**, pp 126-238.

Natural Products in Personal Care and Biomedical Applications

Chapter 7

Synthesis and Characterization of Dual Nanodelivery Systems Containing Vitamin E for Cosmetics and Pharmaceuticals

Rajesh Kumar[1,2,*], Rahul Tyagi[1], Arthur C. Watterson[1], Virinder S. Parmar[1,2], and Jayant Kumar[2]

[1]Institute for Nano-Science Engineering and Technology and Center for Advanced Materials, Departments of Chemistry and Physics, University of Massachusetts, Lowell, MA 01854
[2]Bioorganic Laboratory, Department of Chemistry, University of Delhi, Delhi 110 007, India
[*]Corresponding author: rkumar26@gmail.com

Vitamin E is known for its ability to capture free radicals and therefore to help combat their potential skin damaging effects. Unfortunately it is difficult for it to penetrate and spread through the outer layers of the skin on its own. To get around this problem, dual nano-delivery systems (DND's) having covalently attached vitamin E have been designed and synthesized using novel chemo-enzymatic routes. These synthetic nanocarriers are unique as they have the ability to further encapsulate vitamin E and thus offer high loading and transport capability for vitamin E. Further, these nanocarriers offer both fast and slow release of vitamin E. These nanocarriers also convert highly lipophilic vitamin E to water-soluble form and thus can be used in aqueous preparations for cosmetics and pharmaceutical applications.

There has been a major change in the development of cosmetic products in the recent years. Vitamins and other bioactive substances are increasingly being used alongside conventional ingredients. The immediate cause is increasing knowledge of the 'stress environment' to which the skin, as the human body's largest organ, is exposed. Stress factors include environmental influences such as ultraviolet radiation and the possibly inadequate intake of certain vitamins as a result of modern lifestyles. The present interest in the common roots of health, nutrition and body care is also changing consumer expectations with regard to the performance characteristics of cosmetic products. This development has temporarily pushed the significance of decorative cosmetics into the background. The focus is now on protection and preventive measures against harmful environmental influences and biological aging processes, as well as the possible repair or regeneration of unwelcome changes in the skin. One target area is oxidative stress, *i.e.* the consequences of the creation of reactive forms of oxygen in the skin. The reactive forms that usually undergo physiological metabolization may, under abnormal stress conditions and in the absence of adequate protection by enzymatic and non-enzymatic mechanisms, cause pathophysiological dysfunctions within the skin. Oxidative stress is also an early hallmark of affected neurons in Alzheimer's disease (AD). [1,2] The neurotoxic peptide Amyloid-beta (Abeta), a neurotoxic peptide that accumulates in AD, [3] induces lipid peroxidation, [4,5] which generates downstream cytosolic free radicals and reactive oxygen species (ROS)[6]. Endogenous antioxidant systems normally provide sufficient neuroprotection, but may fail to compensate for the increased generation of reactive oxygen species (ROS) that accompanies neurodegeneration.[7] One limiting factor in dietary consumption of antioxidants may be the difficulty in attaining appropriate concentrations within brain, although recent studies suggest that systemic antioxidant levels are helpful.[8]

Vitamin E is known for its ability to capture free radicals (reactive forms of oxygen) and therefore to help combat their potential skin damaging effects. The antioxidant vitamin E provides limited neuroprotection in AD.[9-12] This limitation is likely to be a consequence of its lipophilic nature, which retards its ability to quench cytosolic ROS, including those generated from antecedent membrane oxidative damage.[6] Pharmaceutical and medical research has for many years focused on vectors as a means of transporting drugs harmlessly to specific parts of the body.[13-16] We have earlier reported a nano-micellar drug delivery system based on the copolymer of poly(ethylene glycol) and dimethyl esters of alkylated phenol, capable of encapsulating lipophilic as well as hydrophilic compounds. [17-20] **Figure 1**, shows the amphiphilic polymer **A** based on poly(ethylene glycol)-co-dimethylester of alkylated phenol and cartoon representation of the micelle formation by the self-assembly of the amphiphilic polymer **A**.

It has been shown earlier that the delivery systems based on the amphiphilic polymer **A** are highly efficient and increase the efficacy of the drug by manifolds when encapsulated as compared to the native form of drug. For example, the use

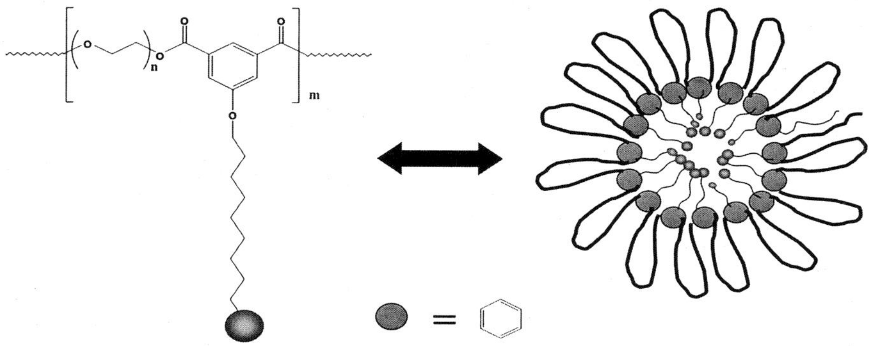

*Figure 1. Cartoon representation of nano-micelle formation by the self-assembly of amphiphilic polymer **A** in aqueous media.* *(See page 11 of color inserts.)*

of these nano-micellar delivery systems appears promising for anti-inflammatory drugs.[18] These nano-micelles are used in particular to transport lipophilic active ingredients, which are soluble in oil. Given that the interstices of the outer layer of skin measure about 100 nanometers, nanocarriers offer the best solution to the problem of transporting and concentrating active ingredients in the skin. In the present work we report a novel dual nano-delivery system (DND's) having covalently attached vitamin E as a way of transporting pure vitamin E.

Experimental

Materials and Methods

Novozyme-435, an immobilized enzyme, was a gift from Novozymes A/S, Denmark. All other chemicals and solvents were of analytical grade and were used as received unless otherwise noted.

Characterization

Gel permeation chromatography (GPC) was used to determine the molecular weights and molecular weight distributions, M_w/M_n of polymer samples. ^{1}H NMR and ^{13}C NMR spectra were recorded on a Bruker Instrument Inc. 250 MHz ARX spectrometer equipped with Silicone Graphics station. Light scattering data was determined by a laser light scattering photometer (Wyatt technology DAWN Model F) equipped with 632 nm He-Ne laser as the light source.

Acylation of Vitamin E with bromoacetyl bromide

To a solution of α-tocopherol (2 g, 4.64 mmol) and triethyl amine (0.7 g, 6.93 mmol) in dichloromethane was added bromoacetyl bromide (1 g, 4.95 mmol) drop by drop at room temperature. The reaction mixture was stirred at room temperature and monitored by thin layer chromatorgraphy (TLC) in 50% hexane/chloroform. After completion of the reaction, the solvent was removed on rotavapour and the residue obtained was redissolved in hexane to remove the salt. The insoluble salt in hexane was filtered off and the filtrate obtained was dried by removing the hexane under reduced pressure, washed with methanol and dried to obtain the desired product **2**.

Coupling of α-tocopherol bromoacetate (2) with Poly [(poly (oxyethylene)-oxy-5-hydroxyisophthaloyl] (3)

Compound **2** (0.86 g, 1.55 mmol) and **3** (1.83 g, 2.32 mmol) were dissolved in dry acetone (20 ml). To this solution was added anhydrous potassium carbonate (0.32 g, 2.32 mmol). The reaction mixture was then refluxed and monitored by TLC in 10 % methanol/chloroform. After completion, the potassium carbonate was removed by simple filtration and solvent was removed under vacuum to obtain the desired product **4**.

Synthesis of the amphiphilic polymer 5

To a solution of **4** (1.5 g, 1.89 mmol) and bromodecane (220 mg, 1.0 mmol) in dry acetone was added anhydrous potassium carbonate and the resulting solution was refluxed in an oil bath maintained at 60 °C. After completion of the reaction (checked by TLC), potassium carbonate was filtered off and acetone was removed to give the amphiphilic polymer **5** as a low melting solid.

Results and Discussion

We have previously reported nano-micelles based on copolymer of poly(ethylene glycol) and dimethyl esters of alkylated phenol **A**, that facilitate intracellular delivery of otherwise nonpermeable molecules.[21] This was accomplished by exposing differentiated SH-SY-5Y cultures to rhodamine either aliquoted directly into culture medium or to an equivalent final concentration of rhodamine that had been encapsulated into nano-micelles described above. No detectable rhodamine fluorescence was observed above background following incubation of cultures with non-encapsulated rhodamine. By contrast, prominent rhodamine fluorescence was observed within the cytosolic compartment of cells incubated with nano-micelle encapsulated rhodamine.[21]

These nano-micelles also found to increase the efficacy of the lipophilic antioxidant vitamin E against Abeta-induced ROS when used in encapsulated form.[21] In continuation of our earlier work and interest on the encapsulation and transport of vitamin E using nano-miceller delivery systems, we report the novel DND's having covalently attached vitamin E as a way of transporting pure vitamin E.

The synthesized nano-carriers are unique due to the flexibility of both bound as well as encapsulated vitamin E. The attachment of the hydrophobic vitamin E in the side chain will help in the micellization in aqueous media and also increase the vitamin E content in the nano-carriers. This also allows a better control over the release and bioavailability of vitamin E and can be used in combination of other bioactive agents. The amphiphilic nature of these nano-carriers make them soluble in both hydrophilic and lipophilic medium. The synthetic approach for these nano-carrriers is shown in **Scheme 1** and **Scheme 2**.

The α-tocopherol bromoacetate **2** was synthesized by the acylation of α-tocopherol with bromoacetyl bromide.[22] The ^{1}H-NMR spectrum of compound **2** showed the signal at δ 4.13 for the $OCOCH_2Br$ confirming the acylation of the phenolic hydroxyl of Vitamin E (**Figure 2**). The α-tocopherol bromoacetate **2** was attached to dimethylisophthalate-co-PEG-600 copolymer **3**, by the alkylation of phenolic hydroxyl in the presence of potassium carbonate to give vitamin E conjugated copolymer **4**. The copolymer **3** was in turn synthesized by enzyme (immobilized *Candida antarctica* lipase B commercially known as Novozyme 435) catalyzed condensation polymerization reaction[17] of dimethyl 5-hydroxy isophthalate with polyethylene glycol (600 g/mol) under solvent less conditions. The alkylation of the phenolic hydroxyl group of **3** by **2** was confirmed by the appearance of a signal at δ 5.06 for the $OCOCH_2O$ protons and disappearance of the signal at δ 4.13 for $OCOCH_2Br$ (**Figure 2**).

The degree of functionalization was determined by comparing the integration of main chain proton at δ 4.50 (for 4 protons) to that at δ 5.06 (for 2 protons) for side chain proton and found to be 30 %. The partial derivatization of phenolic hydroxyl groups was further evident from the two sets of aromatic signals due to substituted and non-substituted aromatic rings. The signal for the substituted aromatic rings appeared lower field as compared to non-substituted aromatic rings. Interestingly, the methylene protons due to the main chain at δ 4.50 also showed splitting, giving two different overlapping triplets (**Figure 2**). These may be due to the different chemical environment of the methylene protons.

The Vitamin E conjugated copolymer **4**, is highly soluble in water but does not aggregate to form nano-micelles as observed by static light scattering technique. This may be due to the improper balance of hydrophilic to hydrophobic segments. Therefore, the partially (30 %) Vitamin E conjugated copolymer **4**, was further reacted with excess of bromodecane to completely alkylate the phenolic hydroxyl group of polymer **4** as shown in **Scheme 2**.

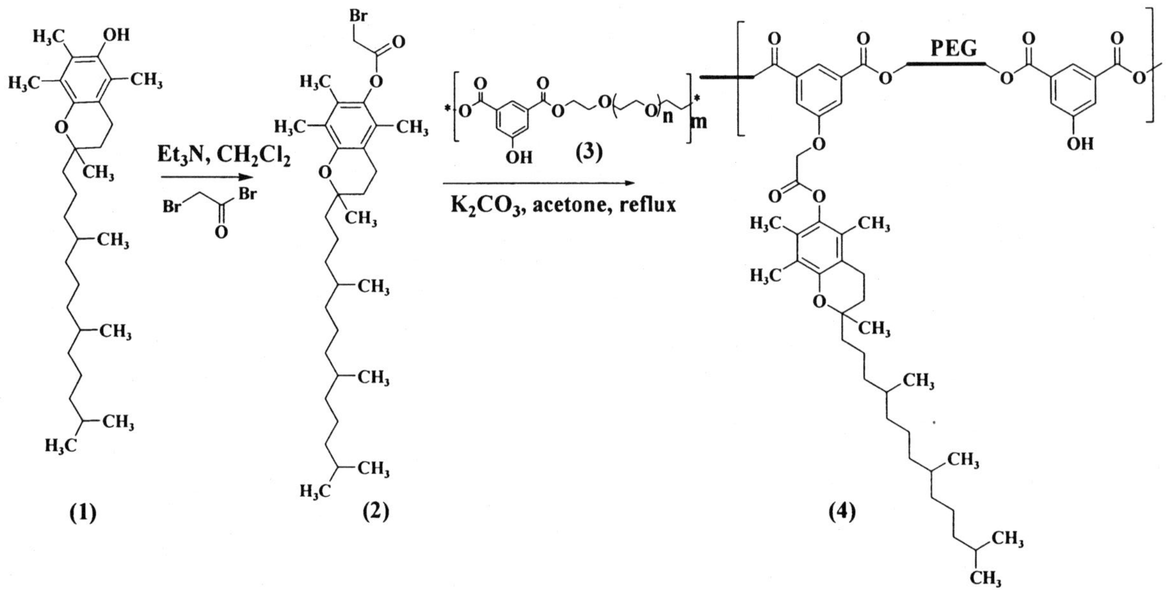

Scheme 1. Synthesis of Vitamin E conjugated copolymer of dimethyl- isophthalate and PEG600, ***4****.*

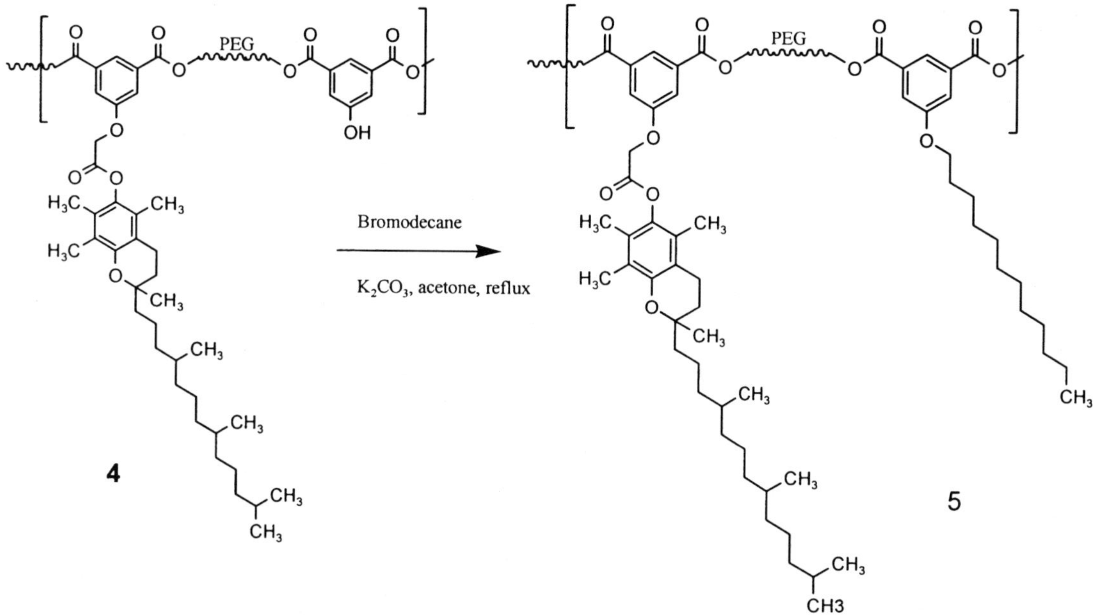

4

5

*Scheme 2. Synthesis of the amphiphilic polymer **5**.*

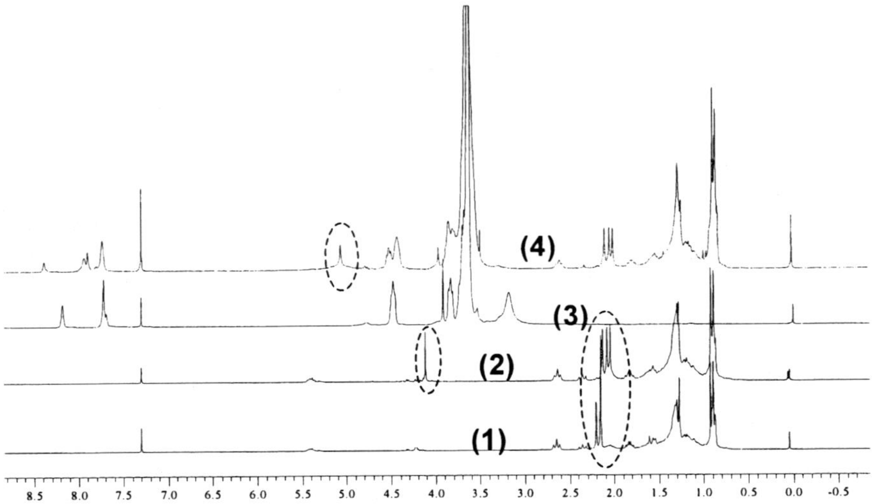

Figure 2. ^{1}H NMR spectra of a-tocopherol (1), bromoacetyl tochopherol 2, copolymer 3, and vitamin E conjugated copolymer 4.

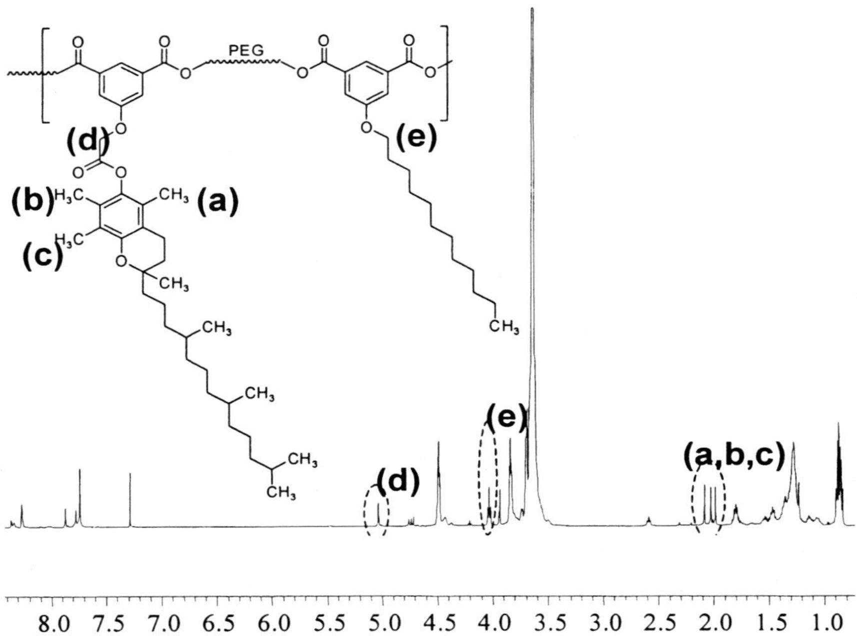

*Figure 3. ^{1}H NMR spectrum of the amphiphilic copolymer **5**.*

The ^{1}H NMR spectrum of the amphiphilic polymer showed a triplet at δ 4.0 and assigned to methylene protons (e) (**Figure 3**). This confirmed the attachment of the bromodecane chain to the phenolic hydroxyl group. The signal at δ 5.06 assigned to methylene protons (d) confirmed the presence of the vitamin E, suggesting that there was no side chain hydrolysis. The composition of the side chains in the polymer **5** were determined by comparing the signal at δ 5.06 for the vitamin E side chain to the signal at δ 4.0 for the decane side chain and found to be 1:2. The aromatic region in the ^{1}H NMR spectrum of the copolymer **5** also showed distribution of the protons due to chemically different environment of the phenyl ring attached to Vitamin E and decane chains substantiating the presence of both side chains, *i.e.* Vitamin E and decane. Surprisingly, the two overlapping sets of triplet at δ 4.50 disappeared and only as expected only one triplet appeared, suggesting the complete substitution of the phenolic hydroxyl groups.

Self-assembly of the synthesized vitamin E containing polymer **5**, in aqueous and organic media was studied by static light scattering. It was observed that when the polymer **5** was dissolved in water, it aggregates to form nano-micelles. The radius of gyration (R_g) and hydrodynamic radius (R_h) of polymer **5** in aqueous solution were determined by static and dynamic light scattering techniques at 25 °C and found to be 57.2 and 10.5 nm. However no micellization was observed in toluene when analyzed by static light scattering. The nano-micelle formed by the polymer **5**, was further used to encapsulate vitamin E and the encapsulation of vitamin E increased from 17 % to 26 % as determined by UV spectroscopy. The nano-micelle formed by the polymer **5** was stable to 65 °C versus 55 °C reported for the nano-micelle formed by the polymer of poly(ethylene glycol)-co-dimetyl ester of alkylated phenol, **A**.[23]

Conclusions

In summary, polymeric nano-carriers containing the bound Vitamin E have been designed and synthesized. The synthesized nano-carriers have the flexibility of further encapsulating Vitamin E or other bioactive agents, and allow a handle for controlling the concentration of Vitamin E in various preparations for cosmetic and pharmaceutical applications. The synthesized Vitamin E containing nano-carriers are highly soluble in both water and oil, making them suitable for both aqueous and non-aqueous preparations. The evaluation of the synthesized nano-carriers in various applications is underway.

Acknowledgements

The authors gratefully acknowledge financial support to RT from CFCI, University of Massachusetts Lowell.

References

1. Markesbery, W. R. and Carney, J.M. **1999**, *9*, 133.
2. Zhu, X.; Raina, A.K.; Lee, H.G.; Casadesus, G.; Smith, M.A.; Perry, G. *Brain Res*. **2004**, *1000*, 32.
3. Huang, X.; Moir, R.D.; Tanzi, R.E.; Bush, A.I.; Rogers, J.T. *Ann N Y Acad Sci*. **2004**, *1012*, 153
4. Koppal, T.; Drake, J.; Butterfield, D.A. *Biochim Biophys Acta* **1999**,*1453*,407.
5. Butterfield, D.A.; Koppal, T.; Subramaniam, R.; Yatin, S. *Rev Neurosci*. **1999**, *10*, 141.
6. Subramaniam, R.; Koppal, T.; Green, M.; Yatin, S.; Jordan, B.; Drake, J.; Butterfield, D.A *Neurochem Res*. **1998**, *23*, 1403.
7. Rottkamp, C.A.; Nunomura, A.; Raina, A.K.; Sayre, L.M.; Perry, G.; Smith, M.A. *Alzheimer Dis Assoc Disord*. **2000**, 14 Suppl 1:S62.
8. Polidori, M.C.; Mattioli, P.; Aldred, S.; Cecchetti, R.; Stahl, W.; Griffiths, H.; Senin, U.; Sies, H.; Mecocci, P. *Dement Geriatr Cogn Disord*. **2004**, *18,* 265.
9. Berman, K.; Brodaty, H. *CNS Drugs*. **2004**, *18,* 807.
10. Growdon, J. *J Alz Dis*.**2001**, *3*, 287.
11. Vina, J.; Lloret, A.; Orti, R.; Alonso, D. *Mol Aspects Med*. **2004**, *25,* 117.
12. Zandi, P.P.; Anthony, J.C.; Khachaturian, A.S.; Stone, S.V.; Gustafson, D.; Tschanz, J.T.; Norton, M.C.; Welsh-Bohmer, K.A.; Breitner, J.C. *Arc Neurol*. **2004**, *61*, 82.
13. Zulli, F.; Suter, F.; Mibelle, A. G.; Buchs, S. *SOFW Journal*. **1997**, *123*, 883.
14. Hildebrand, G.; Niehus, H.; Muller, R. H. *Proc Int Symp Control Rel of Bio Mater*. **1998**, *25*, 433.
15. Mu , L.; Feng, S.S. *J Control Rel*. **2003**, *86*, 33.
16. Mu , L.; Feng, S.S. *J Control Rel*. **2003**, *80*, 129.
17. Kumar, R.; Shakil, N.A.; Chen, M. H.; Parmar, V.S.; Samuelson, L.A.; Kumar, J.; Watterson, A.C. *J Macromol Sci-Pure Appl Chem* **2002**, *39*, 1137.
18. Kumar, R.; Chen, M. H.; Parmar, V. S.; Samuelson, L. A.; Kumar, J.; Nicolosi, R.; Yoganathan, S.; Watterson, A. C. *J Am Chem Soc*. **2004,** *126*, 10640.
19. Kumar, R.; Bruno, F.; Parmar, V. S.; Kumar, J.; Watterson, A. C.; Chittibabu, K. G.; Samuelson, L. A. *Chem Commun*. **2004**, 862.
20. Danprasert, K.; Kumar, R.; Cheng, M.H.; Gupta, P.; Shakil, N. A.; Prasad, A. K.; Parmar, V. S.; Kumar, J.; Samuelson, L. A.; Watterson, A.C. *Eur Poly J*. **2003**, *39*, 1983.
21. Shea, T.B.; Ortiz, D.; Nicolosi, R.J.; Kumar, R.; Watterson, A.C. *J Alz Dis*. **2005**, *7*, 297.
22. Kim, Y.D.; Lee, J.N.; Kim, W.C.; Kim, Y.H.; Kim, M.K.; Ku, M.S.; Cho, I.H. *Ger Offen*. **1999**, DE 19747600.
23. Kumar, R.; Tyagi, R.; Shakil, N.A.; Parmar, V.S.; Kumar, J.; Watterson, A.C. *J Macro Sci-Pure Appl Chem*. **2005**, *42*, 1523.

Chapter 8

Preparation and Characterization of Keratin Coatings for Orthopedic Implant Titanium Rods

Jun Li[1], Chunyong Wu[1], Douglas A. Wicks[1], Robert A. Smith[2], and Sarah E. Morgan[1,*]

[1]School of Polymers and High Performance Materials, The University of Southern Mississippi, 118 College Drive, Hattiesburg, MS 39406
[2]Keraplast Technologies, Ltd., 19206 Huebner Road, Suite 102, San Antonio, TX 78258

Keratin proteins extracted from human hair are employed for developing coatings on titanium. In this work, keratin coatings are prepared on silicon wafer by a two-step reaction: silicon wafers are functionalized with vinyl groups by silanization followed by reaction of keratins with the functionalized surface to form coatings of about 2 nanometers thickness. Through the same protocol, the keratin coatings are prepared on polished titanium rods and blasted rods for use as orthopedic implants. Systematic characterization of coatings on both silicon wafer and titanium is performed by contact angle goniometry, ATR-FTIR, ellipsometry, AFM, EDAX and XPS.

Introduction

Keratins are the major fibrous proteins constituting the outer coverings of the body of humans and animals, including hair, wool, nail and horn. The distinguishing characteristic of keratins is their structural integrity, that is due in large part to the presence of disulfide crosslinks which form a three dimensional network of polypeptide chains. By control of the conditions of disulfide scissio and re-formation, keratins can be isolated from wool or hair and fabricated into various useful forms. Within the past decade, keratins have been developed as biomaterials in multiple configurations, including films, foams, gels, and scaffolds.[1-3] The advantages of keratin proteins as biomaterials include their abundant availability as raw materials, low cost, mechanical strength, and most importantly their unprecedented biocompatibility. Though biomaterials have been made from other natural proteins such as collagen, chitin, chitosan, elastin, gelatin and alginate, they often suffer from some level of incompatibility and exhibit some degree of rejection in human clinical applications. As one of the ubiquitous human proteins, keratin from human hair has intrinsic biocompatibility for use as a biomaterial for medical applications. Specifically, keratin can be easily isolated directly from a patient, thus avoiding immuno-related problems. Another advantage of hair keratin for biomaterials is that keratin is known as the major structural protein in human epithelial cells,[4,5] and plays important roles in wound healing and tissue recovery.[6-8] Many types of keratins in hair are also found to be useful for reepithelization during wound healing.[9-12]

The use of human hair for biomedical applications was first reported by Smith et al.[13,14] Numerous patents have been granted for production of keratin-based products in a variety of forms for potential use in biomedical and personal care applications, including films and sheets,[8] hydrogels,[15] and scaffolds.[16] For example, the use of keratin for bone devices is expected to provide the following potential advantages: does not cause rejection by the body; provides a truly biodegradable implant; is non-toxic and non-irritating; eliminates risk of complication, and eliminates the need for a second operation and thus reduces clinical time, hospital time and recovery time.

The production and use of keratin coatings on titanium for orthopedic implants are described in the patent literature.[17] In-depth characterization of the coatings and coating onto titanium substrates in actual orthopedic implant shapes (such as bone fixation screws), however, are not reported. In this research, keratin coatings are studied by fabrication with and without UV irradiation. Silicon wafer serves as an easy-to-characterize substrate due to its known surface composition and nanoscopically smooth surface. Neat and coated surfaces are analyzed via multiple characterization techniques, including contact angle goniometry, attenuated total reflectance (ATR)-FTIR, atomic force microscopy (AFM) and ellipsometry, to determine optimum coating processes. Finally, keratin coatings are applied to polished and blasted titanium rods for orthopedic implants via optimized coating techniques and characterized using EDAX (energy dispersive x-ray) and XPS (x-ray photoelectron spectroscopy).

Experimental

1. Materials and reagents

The hair used in this study was obtained from a military training facility. Silicon wafers of single side polish were purchased from Silicon Inc. Titanium sheet (0.89 mm in thickness) and rods (3.175 mm in diameter) were purchased from Alfar Aesar. Chlorodimethyl vinyl silane, anthraquinone-2-sulfonic acid sodium salt, hydrogen peroxide (32-35%), and thioglycolic acid were purchased from Aldrich. All other chemicals such as sulfuric acid, DMSO, chloroform, acetone and hexane were purchased from Fisher. All of the reagents were used as received without further treatment.

2. Preparation of Keratins

The keratin isolation procedure followed is that described by Smith et al.[7] Briefly, 60g of clean hair was oxidized by placing in a flask with 1000 mL of 1% hydrogen peroxide to reflux for 3 h. The oxidized hair was separated by filtration, rinsed with copious amounts of water and allowed to air dry. 50 g of the obtained dry and oxidized hair was reduced by placing in 500 mL of 1 mol/L thioglycolic acid at pH 9 (adjusted with ammonium hydroxide) and heating to 60°C under the protection of nitrogen for 24 h. The mixture was subjected to centrifugation, the liquid was collected and added dropwise into an 8-fold volume excess of ethanol to induce precipitation. The precipitated keratin was isolated via filtration and dried under vacuum. The dried filtrate was ground by mortar and pestle to a fine, light grey keratin powder.

3. Keratin coating on silicon wafer and titanium rods

3.1 Silanization of silicon wafer

The silicon wafers were cleaned with freshly prepared concentrated sulfuric acid/hydrogen peroxide (7/3, v/v) for 2 h (caution: the solution reacts violently with organic matter!) and washed extensively with deionized water and acetone.

The cleaned wafers were dried with a stream of nitrogen and used immediately. The cleaned wafers were immersed in a solution of 1% chlorodimethyl vinyl silane in hexanes at room temperature for the designated reaction time and then washed extensively with hexanes and dried.

3.2 Silanization of titanium

Titanium sheet and rods were polished with P4000 SiC paper (5um size) or blasted with #60 grit (254μm size). After ultrasonic cleaning in chloroform, acetone and hexane, they were maintained in D.I. water and dried with N_2 right before use. Silanization of the prepared surfaces was performed according to the procedure described for silicon wafer silanization.

3.3 Formation of keratin coating on silicon wafer and titanium rods

The silanized samples (silicon wafer or polished or blasted titanium rods) were immersed into 5% keratin solution (DMSO/ammonium hydroxide (3N) (90/10 v/v) as the solvent) for the determined reaction time. Coating reactions were conducted both with and without UV irradiation. In the case of UV irradiated samples, a UV lamp (λ=365 nm, 0.01 KJ/cm^{-2}) and 0.5% anthraquinone-2-sulfonic acid sodium salt photo-initiator were employed. After reaction, the substrates were removed from the solution, washed with copious amounts of DI water, submitted to overnight Soxhlet extraction, and dried.

4. Contact angle measurement

After each treatment, advancing water contact angle measurement was performed using the sessile drop technique employing a First Ten Angstroms goniometer (FTA, Portsmouth, VA).

5. Optical ellipsometry

Ellipsometric measurements were carried out on silicon wafers with a Gaertner LSE Stokes Ellipsometer using a He-Ne laser of wavelength of 623.8 nm. The Gaertner Ellipsometer Measurement Program was used to calculate the film thicknesses and refractive indices of both initiator and polymer layers. At least 10 measurements were made and average thickness calculated for each sample. The optical constants of silicon were fixed at Ns = 3.882 and *K*s = -0.019.

6. EDAX(Energy Dispersive X-ray Microanalysis).

EDAX measurements were obtained with an FEI Quanta 200 Environmental Scanning Electron Microscope (ESEM) equipped with a Thermo-Noran Vantage light element energy dispersive x-ray detector, which provides elemental composition down to Beryllium.

7. ATR-FTIR (Attenuated Total Reflectance-FTIR

ATR-FTIR was obtained using a Bruker Tensor 37 equipped with GATR^{TM}-ATR module (Harrick).

8. AFM (Atomic Force Microscopy)

AFM studies were conducted on a Dimension 3000 scanning probe microscope from Digital Instruments (VEECO, Santa Barbara, CA). The images were collected in tapping mode using a standard RETSP silicon cantilever (length: 125 μm; spring constant: 20-80 N/m; and the resonance frequency: about 300 kHz) (VEECO Probes, CA).

9. XPS(X-ray Photoelectron Spectroscopy).

XPS spectra were obtained using a Kratos AXIS 165 X-ray Photoelectron Spectroscope and Scanning Auger Microscope at The Material Characterization Center of Louisiana State University, Baton Rouge, LA

Results and discussion

Silanization of silicon wafer

The silanization of the silicon wafer, the reaction between the hydroxyl groups generated on the silicon wafer surface and the chlorodimethyl vinyl silane, can be monitored by both contact angle measurement and ellipsometry and results in surface coverage with vinyl groups. After silanization, measured advancing water contact angle increases from below 15° (very hydrophilic due to the high concentration of hydroxyl groups) to nearly 70° (moderately hydrophobic due to the formation of the silane layer) (Table I). The effect of reaction time on silane layer thickness as measured by ellipsometry is shown in

Table II. Measured thickness of the silane layer is comparable with the molecular length of the chlorodimethyl vinyl silane molecule, suggesting the formation of a silane monolayer. As thickness does not increase appreciably with reaction time, a silanization time of ten minutes was chosen for further reactions. The formation of vinyl groups on the silicon wafer was also confirmed by ATR-FTIR (Figure 1). In the spectrum of the silanized surface (Figure 1 B), the absorption at 1605 cm^{-1} indicating the presence of vinyl groups (C=C vibration) appears that is not present in the spectrum of the cleaned silicon wafer (Figure 1A).

Table I. Water Contact Angle of Silicon Wafer and Titanium Sheet

Sample	*Contact angle (°)*		
	Neat	*Silanization*	*Keratin coating*
Silicon wafer	14 ± 2	67 ± 3	32 ± 3 (UV for 4h) 33 ± 2 (no UV for 4h)
Titanium sheet	32 ± 2	63 ± 3	35 ± 2

Table II. The Thickness of Silane Layer Determined by Ellipsometry

Reaction time	*10 min*	*1 h*	*24 h*
Thickness (nm)	0.71 ± 0.06	0.76 ± 0.07	0.92 ± 0.09

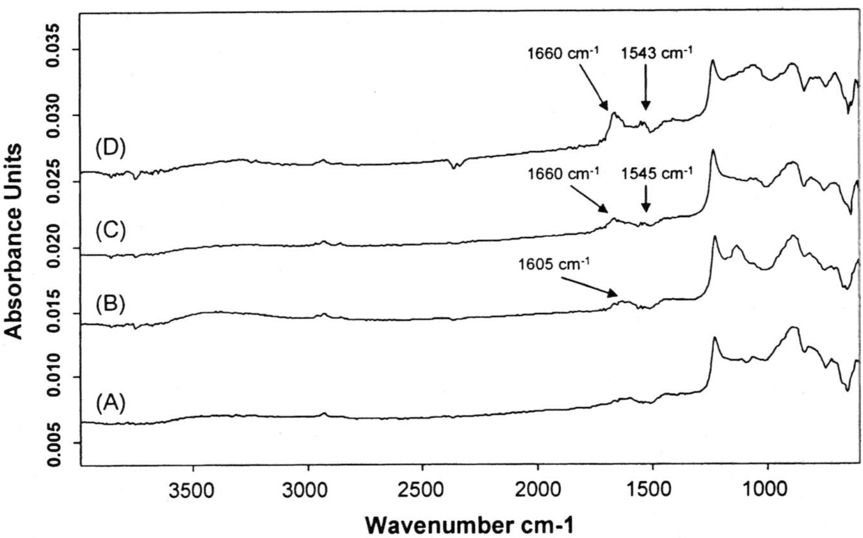

Figure 1. ATR-FTIR spectra of (A) silicon wafer (B) silicon wafer after silanization with chlorodimethyl vinyl silane (C) silalnized silicon wafer reacted with keratin solution with UV for 4h (D) silicon wafer reacted with keratin solution without UV for 4h

Keratin coatings on silicon wafer

The silanized silicon wafer was reacted with keratin solution both with and without UV irradiation and the ATR-FTIR spectra of the obtained samples (after washing and soxhlet extraction) are shown in Figures1C and 1D respectively. Characteristic absorbance peaks at 1660 and 1550 cm^{-1}, attributed to the amide groups in the keratin proteins, appear in both samples. Thus, the ATR-FTIR results indicate the successful formation of keratin coatings on silanized silicon wafers whether UV irradiation was used[17] or not. Additional evidence for keratin coating formation is seen in measured contact angles (Table I), which are reduced to 32 ± 3° due to the formation of the hydrophilic keratin layer above the hydrophobic silane layer on the silicon wafer. Coating thicknesses measured by ellipsometry are given in Table III. In both systems, the keratin coating thickness increases with reaction time over the first four hours, and then reaches a plateau. No clear difference is observed for keratin coating thickness for UV and non-UV irradiated systems. A reaction time of 24 hours was chosen for further experiments.

Table III. Keratin Coating Thickness Measured by Ellipsometry

Sample reaction time (h)		*Thickness (nm)*
UV	0.5	1.21 ± 0.19
	1	1.62 ± 0.24
	2	2.12 ± 0.30
	4	2.22 ± 0.35
	24	2.38 ± 0.33
No UV	0.5	1.28 ± 0.33
	1	1.56 ± 0.28
	2	1.85 ± 0.21
	4	2.10 ± 0.23
	24	2.29 ± 0.18

AFM studies were performed on neat, silanized and keratin coated silicon wafers (Figure 2). Spherical raised structures and increased surface roughness are apparent in the silanized sample (Figure 2B) in comparison to the cleaned silcon wafer (Figure 2A). The surface roughness further increases for the keratin coated wafer, and a more continuous coating appears (Figure 2C). The AFM results further confirm the silanization of the silicon wafer and subsequent formation of the keratin coating.

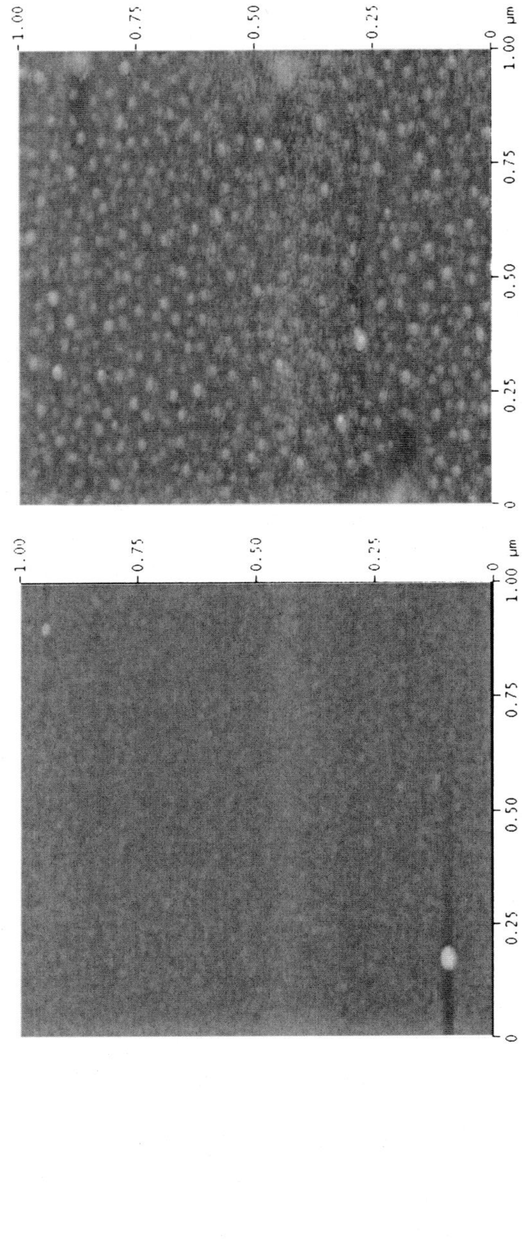

2(A) 2(B)

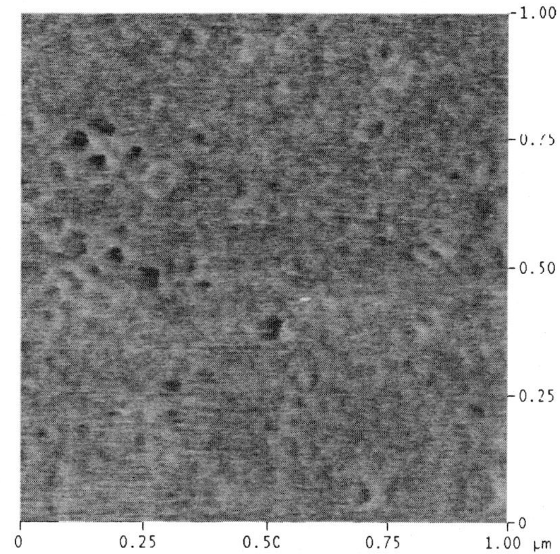

2(C)

Figure 2. The AFM image of silicon wafers (A) clean (RMS = 0.167 nm); (B) after silanization (RMS = 0.437); (C) with keratin coating (4h, without UV irradiation) (RMS = 0.687 nm) degradation of the keratin coating. Discoloration and changes in IR spectra observed for UV-irradiated samples (especially below 1200 cm^{-1}, Figure 1) provide evidence of these unexpected reactions for samples with UV exposure. Based on these results, keratin coatings were prepared on titanium substrates following the same protocol used for silicon wafers, without UV irradiation.

(See page 12 of color inserts.)

The above results indicate that keratin coating is formed on immersion of the silanized surface in keratin solution without the use of UV irradiation or UV initiator. Furthermore, extended exposure of the materials to UV irradiation may lead to formation of other reactive species, further complex reactions and

Silanization of titanium

Model reactions were performed on titanium sheets. To simulate actual implants, reactions were performed on titanium rods (polished and blasted). The silanization reaction for titanium samples was identical to that used for silicon wafers. The successful silanization of titanium is evidenced by contact angle goniometer measurements (titanium sheets) and EDAX. Advancing water contact angle increases from 32±2° for cleaned sheet to 63±3° for silanized sheet (Table I). As was observed for the silicon wafer reaction, increased reaction time up to 24 hours does not lead to further increase in contact angle. EDAX studies indicate successful formation of the silane layer on titanium sheet by the appearance of the silicon peak (Figure 3), not present in the neat titanium spectrum. It should be noted that cleaned titanium samples were immersed in deionized water until use. After drying, they were immediately immersed in silane solution, to avoid inevitable oxidation that would occur on exposure to atmosphere that would interfere with silanization reactions.

Keratin coatings on titanium samples

Keratin coatings were prepared on titanium samples under the same conditions used for silicon wafers, 24 hour immersion in keratin solution without UV exposure. After keratin reaction on titanium sheets, measured contact angles are reduced to 35 ± 2°, in the same range as those measured for keratin coated silicon wafers (Table I). For the titanium rods (non-flat surfaces) XPS was employed to determine keratin coating thickness. The signals of elements of Ti, Si and N were measured via XPS for neat, silanized and keratin coated titanium rods, both polished and blasted. For example, the change in the intensity of the Ti peak for polished rods and blasted rods after each step are given in Fig.4A and Fig. 4B respectively. In both cases, the Ti signal intensity decreases with silanization and further decreases with keratin coating, indicating the presence of the coatings. Ti concentration on the surface is expected to decrease as the coating thickness increases. Ti signal intensity is weaker for the blasted titanium rod than for the polished rod, indicating a thicker coating for the blasted sample.

The thickness of keratin coating is calculated based on the attenuation of the Ti signal intensity after each step of coating using the following equation:

$$N=N_0 \exp[-t/(\lambda * \sin\theta)]$$ [18]

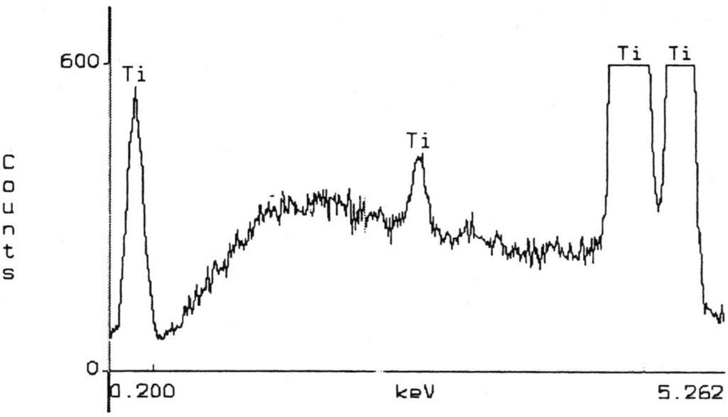

3(A)

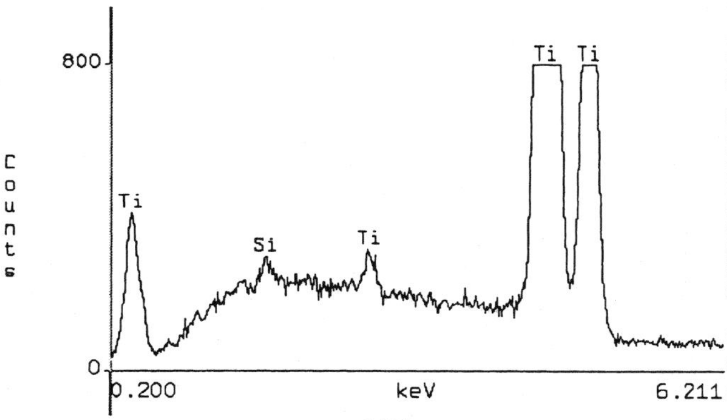

3(B)

Figure 3. EDAX spectrum of the blasted titanium sheet before (3A) and after silanization (3B)

Where:
N_0: Ti intensity of blank rod;
Ni: Ti intensity of coated rod;
λ: free path of photoelectron (assumed 20Å);
θ: the angle between detector and rod surface (θ is 90 degree here).

The calculated thicknesses of silane and keratin coating layers on polished rods and blasted rods are shown in Table IV. The blasted rods yield substantially thicker coatings, presumably due to the inherent roughness of the blasted surfaces, which allows pooling of the coating material during the immersion process. Additionally, coating thickness was measured after overnight soxhlet extraction of the coated rods. These measurements show the stability of the coating, with only minimal reduction in thickness after overnight extraction.

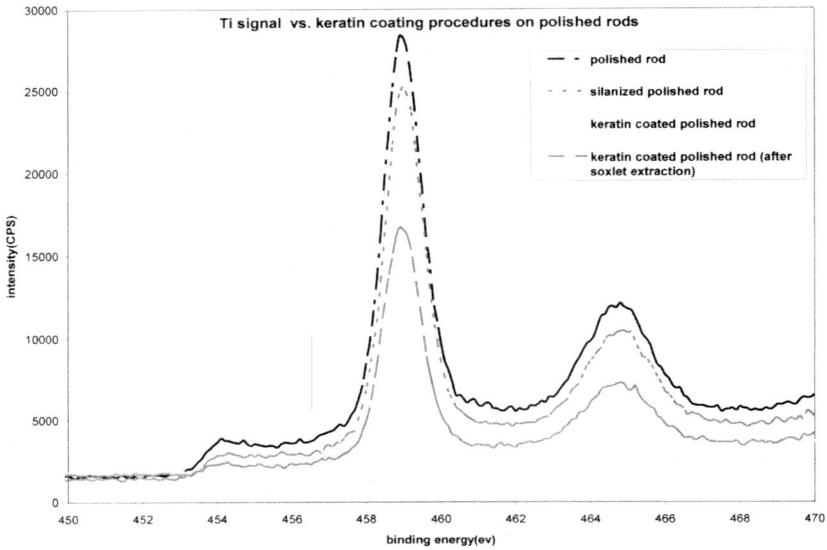

4(A)

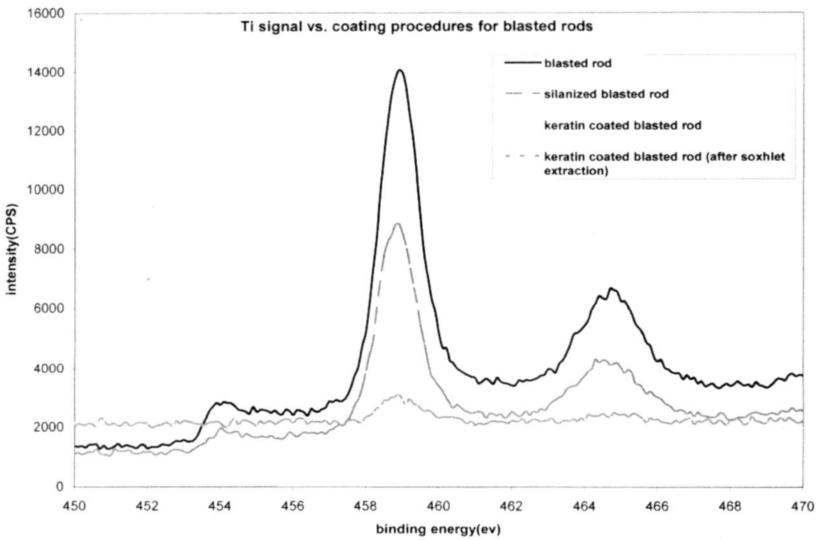

4(B)

Figure 4. XPS spectra of Ti for polished (A) and blasted (B) titanium rods after each step of treatment
(See page 13 of color inserts.)

Table IV. Estimated Coating Thickness on Titanium Rods by XPS

	Thickness (nm)	
	Polished rods	*Blasted rods*
silane	0.403	0.950
Keratin coating	1.584	7.770
Keratin coating after soxhlet extraction	1.103	4.354

Conclusions

Stable keratin coatings of approximately two nanometers thickness were successfully prepared on titanium rods for orthopedic implants. Both polished and blasted titanium rods were successfully coated, and optimized coating techniques were developed. It was found that UV irradiation is not necessary for the formation of stable coatings. Extensive characterization of model substrates and titanium rods provides evidence for coating thickness and stability.

References

1. Tanabe, T.; Okitsu, N.; Yamaguchi, K. *Mater. Sci. Eng. C: Biomimetic Supramole. Sys.* **2004**, *C24*, 441-446.
2. Yamaguchi, K.; Hojo, H.; Yamamoto, Y.; Tanabe, T. *Mater. Sci. Eng. C: Biomimetic Supramole. Sys.* **2004**, *C24*, 467-472.
3. Tachibana, A.; Kaneko, S.; Tanabe, T.; Yamauchi, K. *Biomaterials*, **2005**, *26*, 297-302.
4. Coulombe, P. A. ; Tong, X. , Mazzalupo, S.; Wang, Z.; Wong, P. *Eur. J. Cell. Biol*, **2004**, *83*, 735-746.
5. Kirfel, J.; Magin, T.M.; Reichelt, J. *Cell. Mol. Life Sci.* **2003**, *60*, 56-71.
6. Smith, R. A.; Blanchard, C. R.; Lankford, J. Jr. S. F.; U.S. Patent 5,358,935, 1994
7. Timmons, S. F.; Blanchard, C. R.; Smith, R. A. U.S. Patent 5,948,432, 1999.
8. Timmons, S. F.; Blanchard, C. R.; Smith, R. A. U.S. Patent 6,124,265, 2000.
9. Coulombe, P. A., *Prog. In Dermatology*, **2003**, *37*, 219-230.
10. Bruen, K.J.; Campbell, C.A.; Schooler, W.G.; de Serres, S., *J. Surg. Res.* **2004**, *120*, 12-20.
11. McGowan, K. M.; Coulombe, P. A. *J. Cell Biol.* **1998**, *143*, 469-486.
12. Paladini, R. D.; Takahashi, K.; Bravo, N.; Coulombe, P. A., J. Cell Biol. **1996**, *132*, 381-97.

13. Blanchard, C. R.; Timmons, S. F.; Smith, R. A. U.S. Patent 6,159,496, 2000.
14. Blanchard, C. R.; Timmons, S. F.; Smith, R. A. U.S. Patent 6,379,690, 2002.
15. Blanchard, C. R.; Timmons, S. F.; Smith, R. A. U.S. Patent 5,932,552, 1999.
16. Timmons, S. F.; Blanchard, C. R.; Smith, R. A. U.S. Patent 6,432,435, 2002.
17. Van Dyke, M.; Siller-Jackson, A. J. U.S. Patent 20050058686, 2005.
18. Fadley, C. S. *Prog. Surf. Sci.* **1984**, *16*, 275-388.

Silicone Polymers in Cosmetics

Chapter 9

Silicone Polymers: New Possibilities in Nanotechnology

Anthony J. O'Lenick, Jr.

Research and Development, Siltech LLC, 2170 Luke Edwards Road, Dacula, GA 30019

According to the National Science and Technology Council Committee on Technology Interagency Working Group on Nanoscience, Nanotechnology has been defined as the creation and utilization of materials, devices, and systems through the control of matter on the nanometer-length scale – that is, at the level of atoms, molecules, and supramolecular structures. Silicone technology offers the ability to make very specific molecules through the control of molecules. This makes them of great interest in the field of nanotechnology.

Silicone polymers and their ability to form self assembling units is the key to the functionality of these materials in making useful nanostructures. The first step is to engineer the polymer, using well known techniques. The next step is to look at the functionality of these materials in the solvent system of interest. The driving force for assembly is obtaining the lowest free energy in the system. Many times the lowest free energy state is not the ease, but rather the most ordered. This is particularly true in aqueous systems. The reason for this is that breaking up on hydrogen bonding in water is a very costly process energetically. Large molecules will assume many structures that are more energetically favorable than breaking up many intra-molecular hydrogen in water. The silicone polymer is interesting since internally substituted materials (the so-called comb materials) are free to rotate around the Si-O-Si backbone to obtain the lowest free energy system. The so-called terminal substituted silicones do not have free rotation available forming more complicated hair in and other larger structures.

Silicone Technology

Silicone compounds have been known since the 1860, but it was not until the pioneering of Rochow in the 1940s that this important class of compounds achieved commercial viability. This was due in large part to the development of a process which was called the direct process, and now bears Rochow's name. Silicone chemistry provides the polymer chemist with the ability to construct precise molecules having desirable nanotechnology properties.

The Rochow process[1] is based upon the reaction of silicon metal and methyl chloride to make a series of products called chlorosilanes. The reaction is as follows;

$$
\begin{aligned}
Si + CH_3Cl \rightarrow\ & (CH_3)_4\text{-}Si \\
& (CH_3)_3\text{-}Si\text{-}Cl \\
& (CH_3)_2\text{-}Si\text{-}(Cl)_2 \\
& CH_3\text{-}Si\text{-}(Cl)_3 \\
& Si\text{-}(Cl)_4 \\
& CH_3\text{-}SiH(Cl)_2 \\
& (CH_3)_2\text{-}SiHCl
\end{aligned}
$$

The preparation of chlorosilanes is practiced by a small number of basic manufacturers that grind up silicon metal and react it in a tubular reactor with methyl chloride. The manufacturers of chlorosilanes are referred to as "crushers".

Chlorosilanes are hydrolyzed in water to make intermediates used to make silicone derivatives. The reaction product of water and chlorosilanes is referred to as hydrolyzate.

The chlorosilanes are placed into water, HCl stripped off and after distillation and a variety of clean up processes, a series of silicone building blocks emerge. The most important of which include; hexamethyldisloxane (MM), cyclomethicone (D4) and silanic hydrogen compound. That is just the start of the story. These materials are combined to make compounds of interest using three types of processes. These are[2]:

A. Construction
B. Functionalization
C. Derivatization

A. Construction

Highly specialized activities often create a jargon or language that makes facilitates improved more rapid communication to its practitioners, and keeps people outside the filed from feeling comfortable in these specialized activities. Chemistry, law and government are but a few examples. Silicone chemistry is also an example[3]. The language makes use of the letters M, D, T and Q to specify structural groups placed into a molecule by its construction. The construction step is the process in which the length of the polymer chain, its branching and its positions for insertion of organic groups is determined. The shorthand for construction is as follows;

CH_3
|
$CH_3\text{-Si-}O_{1/2}\text{-}$ M for mono functional with regard to oxygen
|
CH_3

CH_3
|
$\text{-}O_{1/2}\text{-Si-}O_{1/2}\text{ -}$ D for di functional with regard to oxygen
|
CH_3

$\text{-}O_{1/2}$
|
$\text{-}O_{1/2}\text{-Si-}O_{1/2}\text{-}$ T for tri functional with regard to oxygen
|
CH_3

$\text{-}O_{1/2}$
|
$\text{-}O_{1/2}\text{-Si-}O_{1/2}\text{-}$ Q for quad functional with regard to oxygen
|
$O_{1/2}\text{-}$

Clearly, there are no real ½ O. This nomenclature is used so that when two or more groups are linked together a single oxygen exists between them. For example MM (hexamethyldisiloxane) a key material is often referred to as 0.65 viscosity silicone fluid or hexamethyl disiloxane. Its structure is:

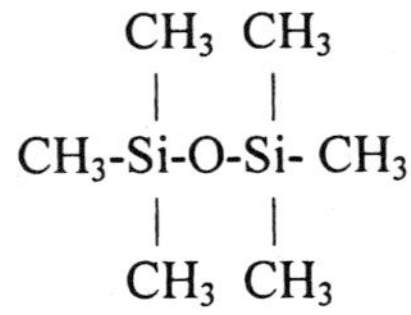

M units are chain terminators since they are mono-functional.

Dimethicone[3]

Reaction of M and D will result in silicone fluids. M units are chain terminators and D units are linear chain extenders. They provide larger and larger molecules having higher and higher viscosity. Historically, silicone fluids are the most understood and commonly used silicone product. These materials are homopolymers that are insoluble in water and in mineral oil. Dimethicone is also called by a number of less technically acceptable terms including silicone fluids, silicone oils, or merely silicones. They are sold by viscosity and vary from low molecular weight very thin materials to products that are very thick and sticky.

Group Opposites[4]

So far all compounds discussed are homo-polymers of silicone. They are water insoluble making the hydrophobic, but they are also oil insoluble. Normally, materials that are hydrophobic contain oil loving groups (referred to as oleophillic groups). However, in this case the water insoluble group is oil hating (oleophobic) and silicone loving (siliphilic). This complexity has resulted in the introduction of the concept of group opposites.

Hydrophobic (Water Hating) can be either Siliphillic (silicone loving) or oleophillic (oil loving).

Siliphillic (silicone loving) is both oleophobic (oil hating) and (water hating) hydrophobic (water hating).

Oleophillic (oil loving) is both siliphobic (silicone hating) and hydrophobic (water hating)

One application for this concept is to coated pigments. Almost all pigments have some sort of coating on them. An oil coating or a silicone coating. The ability to disperse the pigment efficiently is achieved using the phase in which the coating is most compatible. Consequently, a silicone coated pigment often gets used in a silicone phase. An oil coated pigment often gets used in an oil phase. One additional comment n pigments, some are chemically reacted forming covalent bonds between pigment and coating. Others are merely chemisorbed. Those pigments in which the coating is not chemically bonded can be metastable in emulsion systems. Keeping in mind that the materials in an emulsion will go to the phase in which the lowest free energy is achieved; the non-bonded pigment can over time migrate off the pigment into another phase. The result can appear as emulsion instability over time. The modification of the emulsifier package will not solve this problem. We recommend testing all pigments for the type of coating and its permanence.

B. Functionalization

Up to now we have only considered silicone homo-polymers. This class is best understood and an important class of compounds, but only a small portion of the total products available to make products useful in the personal care market. It would indeed be a sad situation if the organo-functional materials were not available, or if the formulation chemist was not made aware of the advantageous of such materials.

The preparation of a silanic hydrogen containing polymer by the construction process does not alter solubility. The silanic hydrogen pre-polymer assumes its altered solubility only after the functionalization reaction is run. For this reason silanic hydrogen containing polymers are considered precursors to organo-functional products. A single silanic hydrogen polymer can give rise to an entire family of analogs depending upon which functional group is placed onto the backbone in the functionalization reaction.

In order to make these products more easily formulated, organo-functional dimethicone compounds have been developed. These include dimethicone compounds with improved oil soluble called alkyl dimethicone compounds; dimethicone compounds with improved water solubility, called PEG/PPG dimethicone. There are also a series of compounds in which surfactant groups are grafted onto the backbone to improve virtually all surfactant properties including detergency, conditioning, wetting, and emulsification. This ability to provide silicone products with improved applicability in personal care products, not only opens the possibility of many high performance products, but also can be a source of frustration to many formulators whom have not been given the necessary structure / function relationships to make intelligent choices in picking products. Often the formulator is left to use products recommended by suppliers, rather than to be a participant in choosing the optimized product for an application. The key to avoid this situation is to learn the rules of structure / functions related to silicones and apply them to new products, resulting in the most cost effective products possible. This article will review those important relationships and propose compounds for the formulator to consider.

The reaction used to place organo-functionality into silicone compounds is called hydrosilylation. This process is used in the construction part of silicone preparation.

The key reaction is one in which a silanic hydrogen (Si-H) is reacted with a terminal double bond. This results in a stable Si-C bond.

$$\begin{array}{l} \backslash \\ \text{-Si-H} \end{array} + CH_2{=}CH\text{-}(CH_2)_7CH_3 \rightarrow \begin{array}{l} \backslash \\ \text{-Si-}(CH_2)_9CH_3 \\ / \end{array}$$

The shorthand for the construction of reactive compound is as follows;

$$\begin{array}{c} CH_3 \\ | \\ CH_3\text{-Si-}O_{1/2}\text{-} \\ | \\ H \end{array}$$

M* for mono functional with regard to oxygen and a reactive H

$$\begin{array}{c} CH_3 \\ | \\ \text{-}O_{1/2}\text{-Si-}O_{1/2}\text{-} \\ | \\ H \end{array}$$

D* for di functional with regard to oxygen and a reactive H.

$$\begin{array}{c} \text{-}O_{1/2} \\ | \\ \text{-}O_{1/2}\text{-Si-}O_{1/2}\text{–} \\ | \\ H \end{array}$$

T*for tri functional with regard to oxygen and a reactive H.

These materials are reacted with in the equilibration reaction to make reactive intermediates which are hydrosilylated in the functionalization reaction. The vinyl containing groups that are reacted include;

a. Alpha olefin $CH_2{=}CH\text{-}(CH_2)_7CH_3$
b. Ally alcohol alkoxylates $CH_2{=}CH\text{-}CH_2\text{-}O(CH_2CH_2O)_8H$
c. Fluoro vinyl compounds $CH_2{=}CH\text{-}CH_2\ (CF_2)CF_3$

The comparison of products within a class shows the importance of the construction.

PEG-8 Dimethicone Water Soluble Silicone[5]

The molecular weight of a PEG 8 dimethicone has a dramatic effect upon performance of the compound as far as both wetting and irritation. Molecular weight is a measure of construction. The graph below shows the molecular weight of molecules studied. Essentially, in the construction the compounds have the same ratio of D to D*, only higher numbers of both.

Wetting is a critical often overlooked aspect of cosmetic formulation. If you are applying something to the air, skin, a pigment or any other surface how that surface wets is critical to the usefulness of the formulation. Low concentration of silicone wetting agent (0.1 to 1.0 % by weight) needs to be added to make the product function correctly. Construction also has a profound effect upon eye irritation, essentially disappearing at a molecular weight of 2412.

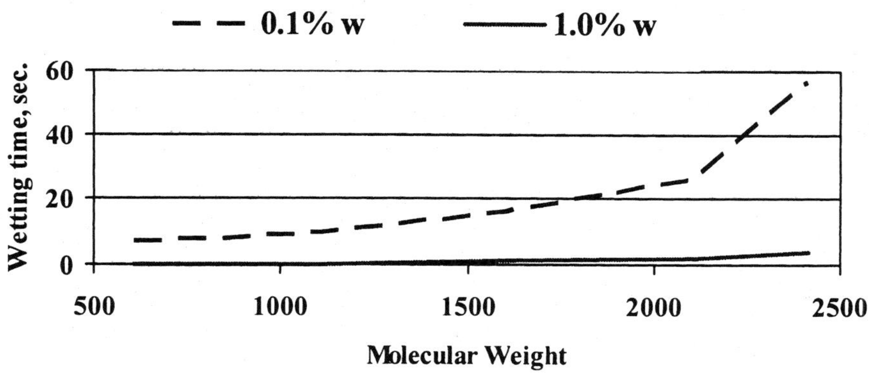

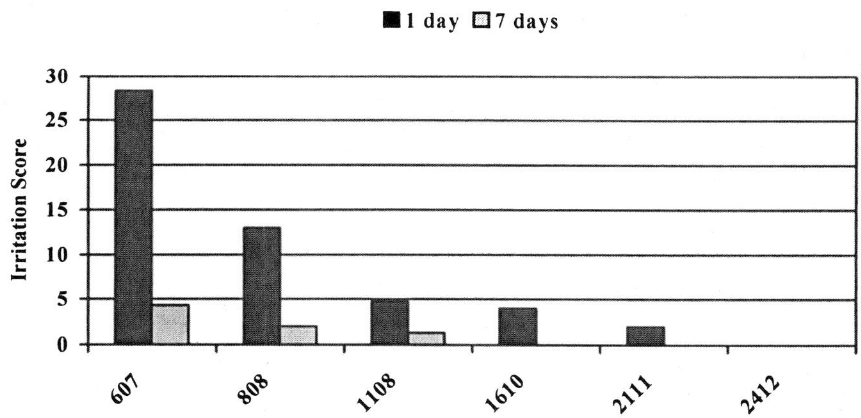

As the molecular weight increases, silicone compounds with the identical amount of PEG, exact same ratio of D to D*, the products transform from wetting agents to emulsifiers, to conditioners to water proofing agents. This transformation is a direct consequence of the lowest free energy conformation the molecule assumed in water. This is in turn related to the ease at which rotation occurs around the molecule backbone.

Alkyl Dimethicone Oil Soluble Silicones[6]

Just as allyl alcohol alkoxylates can be reacted with silanic hydrogen polymers to giving products with improved water solubility, they can also be reacted with alpha olefin to produce a series of oil soluble silicone polymers with very interesting properties. Silanic hydrogen compounds can be reacted with alpha olefin, to give mineral oil soluble materials. The compounds conform to the following structure;

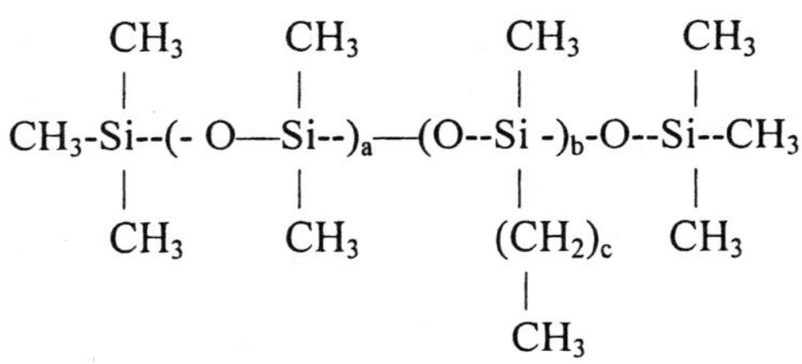

The inclusion of a very small amount of alkyl group into the molecule results in a product that is soluble in mineral oil. This solubility offers the ability to use these materials in a variety of clear systems based upon many oils, where the use of silicone fluids in these systems is not possible,

There are several important trends that become apparent when one looks for a structure / property relationship. They include;

1. The length of the alkyl chain has a dramatic effect upon the melt point of the wax. Waxes based upon alkyl groups having between 12 and 16 carbon units are liquids at room temperature. At carbon lengths of 18 and above products become solid and the melt point increases as the carbon length goes up.

2. The amount of silicone in the molecule, for a given alkyl length, has a minimal impact on melting point.

3. The amount of silicone in the molecule, for a given alkyl length has a minimal, has an effect upon wax hardness. As the amount of alkyl group in a wax is increased in a wax having over 18 carbon atoms in the alkyl group, the wax becomes harder.

Solubility (10% Weight)

Water	Mineral oil	PG	D-5	Sil Flud 350 Visc	Iso propanol	Aromatic Hydrocarbon
Cerotyl Dimethicone (Silwax D-026)						
I	S	I	D	D	I	D
Cerotyl Dimethicone (Silwax J-226)						
I	S	S	I	D	I	S
Stearyl Dimethicone (Silwax H-418)						
I	S	S	I	D	I	S
Stearyl Dimethicone (Silwax L-118)						
I	S	S	I	I	I	S

Legend I = insoluble d= dispersible s = soluble

Physical Properties of Silicone Waxes having different amounts of Alkyl Group on same Silicone Backbone

Silwax	Alkyl Group	State RT	% Silicone	% Alkyl	MP (°C)
J1012	Lauryl	Liquid	82.0	18.0	Liquid (RT)
J1016	Cetyl	Liquid	77.0	23.0	Liquid (RT)
J1018	Stearyl	Liquid	75.0	25.0	Liquid (RT)
J1022	Behenyl	Soft solid	72.0	28.0	20.0
J1026	Cerotyl	Solid	68.0	32.0	46.0
J1032		Hard solid	64.0	36.0	60.5

Physical Properties of Silicone Waxes having different amounts of Silicone using the same Alkyl Group

Silwax	Alkyl Group	State RT	% Silicone	% Alkyl	MP (°C)
D1026	Cerotyl	Solid	69.0	31.0	42.0
D2026	Cerotyl	Solid	81.0	19.0	37.0
D3026	Cerotyl	Solid	86.0	14.0	36.0
H-418	Stearyl	Liquid	61.0	39.0	Liquid (RT)
P-418	Stearyl	Soft Solid	58.0	42.0	Liquid (RT)
L-118	Stearyl	Soft Solid	38.0	62.0	30.0

Understanding these trends allow for the selection of a wax for the specific application chosen. All silicone waxes offer improved oil solubility over silicone fluids. Waxes added to oil phases offer an ability to alter the viscosity and skin feel of a formulation. Mineral oil can be gelled by addition of the proper wax. Petrolatum can be thinned out and made less grainy by adding liquid waxes. The play time at a given melt point can be altered by selecting the specific alkyl chain (melt point) and adding differing amounts of silicone to the molecule (hardness).

In addition, silicone waxes are great at minimizing syneresis in pigmented products. The alteration of the amount of alkyl group relative to silicone group determines which wax is best at lowering syneresis in a given formula.

C. Derivatization

"Derivatization" relates to the chemistries practiced on the groups added to the silicone backbone by functionalization. An example of a compound that has construction, functionalization and derivatization is PEG-8-dimethicone phosphate.

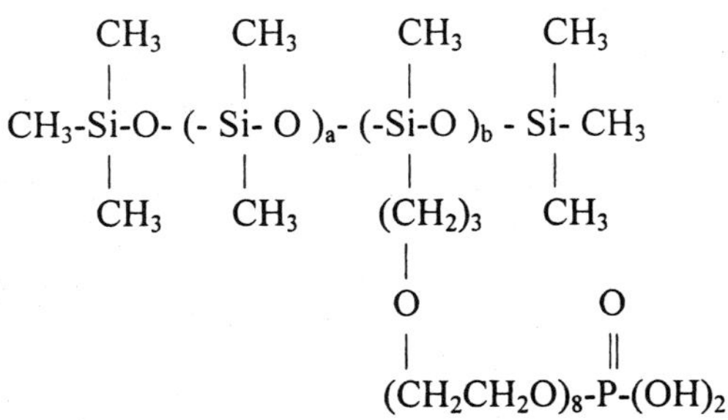

United States Patent 5,859,161 to Imperante issued January 12, 1999 teaches that silicone phosphate esters of the type shown above mitigate eye and skin irritation when used in synergistic combinations with fatty alcohol sulfates and fatty alcohol ether sulfates. These compounds are also used in sunscreens to improve the deposition of the film on the skin and improve SPF values.

Conclusion

Silicone compounds offer a wide range of possibilities to engineer molecules that offer nanostructures in a variety of applications. The understanding of the chemistry is key to effective use of the products. The properties of these materials in various solutions vary as the structure does. The making of several types of compounds and the study of their properties is key to efficient use of these materials. It requires candor on the art of the silicone manufacturer and evaluation on the part of the formulator to be successful. The possibilities offered to the formulation chemist by silicone chemistry is endless and yours to explore.

References

1. O'Lenick Jr., Anthony J. *Silicones for Personal Care*, Allured Publishing Corporation (2003), p. 13.

2. O'Lenick Jr, Anthony J. *Silicones for Personal Care*, Allured Publishing Corporation (2003), p. 35.

3. Stock, Alfred. *Ber. Deutsch. Chem Ges,* (1916), 49:108

4. O'Lenick Jr, Anthony J., JSD 2,229 (2000), p. 387.

5. O'Lenick Jr, Anthony J. *Silicones for Personal Care*, Allured Publishing Corporation 2003, p. 59.

6. O'Lenick Jr, Anthony J. *Silicones for Personal Care*, Allured Publishing Corporation 2003, p. 49.

7. O'Lenick Jr, Anthony J. *Silicones for Personal Care*, Allured Publishing Corporation 2003, p. 75.

Chapter 10

Silicones in Personal Care Applications

Roy U. Rojas Wahl[1,*], J. R. Nicholson[1], and Judith L. Kerschner[1,2]

[1]GE Advanced Materials, Silicones, 771 Old Saw Mill River Road, Tarrytown, NY 10591
[2]International Flavor and Fragrances, Inc., 1515 State Highway #36, Union Beach, NJ 07735
[*]Corresponding author: roy.rojas-wahl@ge.com

Silicone fluids and modified silicone fluids can provide a range of benefits when formulated in various shower/bath products. The nature of these benefits is dependent upon the functionality on the silicone polymer and on the properties of the shower or bath product into which they are introduced. Thus, different silicones can be used to modify in-use product properties, enhance lather, and deliver a variety of post-use sensory experiences. Some silicone materials can assist in deposition from wash-off products. Others can be blended with organic oils, such as triglyceride oils, petrolatum, or mineral oil, to provide a novel after-wash feel while maintaining the occlusive or moisturizing properties of the organic material. To optimize the benefits of these silicones in shower/bath products, it is critical to understand the interactions of the silicone materials with the other standard formulation ingredients such as surfactants, emulsifiers, perfumes and polymers and how these interactions will affect the product performance. This work will highlight some of the key benefits that silicone fluids can deliver from a range of bath/shower product formats.

Introduction

Body washes and shower gels account for about 56 % of the total of the bath and shower sector in the U. S.[1] Traditionally, the use of silicone fluids in such personal washing applications has been limited to dimethicone oils, often for conditioning purposes. There remain, however, many unmet formulation needs such as actives or fragrance deposition, since much of those ingredients tend to be just washed away and are given only a very limited amount of time to approach or penetrate the skin. Also, enhanced moisturization with superior sensory properties is of high importance. These two and more arguments hold the need for new and innovative technologies. Here we show that multifunctional silicones can deliver perceivable benefits, including those aforementioned, and additionally, modify lather and deliver a variety of post-use sensory experiences. First, we introduce the unique characteristics of several polyether copolyols. Then we focus on head-space measurements for the investigation of fragrance deposition, which is enhanced by a combination of alkyl modified silicone waxes and high viscosity dimethyl fluids, showing unique synergistic benefits. Also, we show how the tackiness of a classic occlusive material such as petrolatum can be migitated through the use of a superspreading trisiloxane without compromising transepidermal water loss, and we introduce sensory enhancing silicone crosspolymer gels and their effect on important sensory attributes such as tackiness, oiliness, silkiness and afterfeel.

Experimental

Fragrance Measurements (See Figure 1). Over Product: 500 mg of body wash product was added to a 20mL GC vial and left to equilibrate for 24 hours. Then samples were analyzed by GC-MS using Solid Phase Microextraction (SPME) sampling with an Agilent 6890 GC with 5973 MS Inert detector equipped with a Gerstel MPS-2 autosampler.

Over Skin[2,3]: 1.0 g of body wash was applied to a wet volar forearm and the arm was washed for 30 seconds, rinsed for 10 seconds and patted dry. A specially designed round-bottomed flask equipped with a "sealed cap" was fitted over the arm and an SPME needle was inserted into the flask. The headspace above the arm was sampled for 30 minutes and then the SPME needle was injected into an Agilent 5890 GC with a 5970 MDS detector. The total fragrance areas were then calculated and compared across products (Figures 5 & 6).

Sensory Measurements. A 12 person naïve internal panel was asked to apply a small amount of material on the forearm and then rank 5 different skin feel sensory attributes on a scale of 1-10. Sunflower Oil was used as a standard and set to a score of 5 on each attribute. The panelists then scored the other

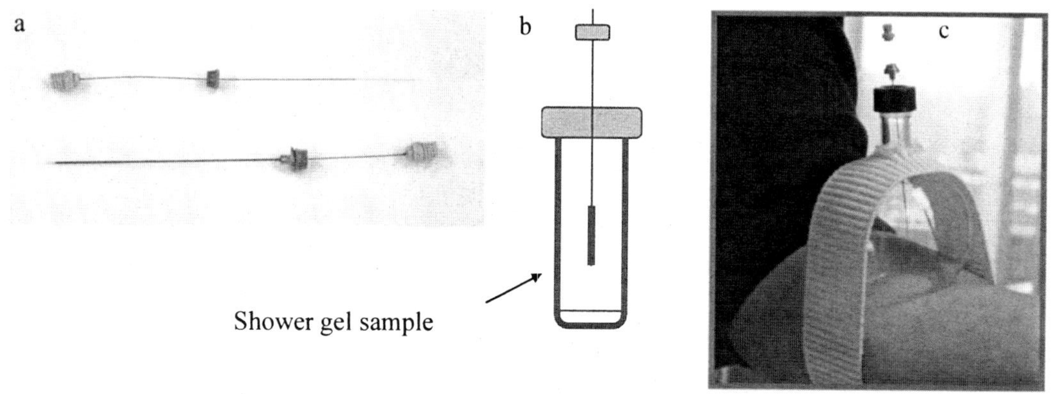

Figure 1: a) SPME needles coated with PDMS/Carboxen to absorb volatiles from atmosphere. b) Sampling from product itself. c) Sampling of deposition on skin after wash

products based on the standard (i.e. less tacky than sunflower oil, score <5). The scores from each product were averaged and recorded in Figure 9.

Skin Water Loss Measurements. Vitro Skin™ was cut into 1.5x6.5 cm rectangles and hydrated for 24 hours (glycerol/water). Each piece was mounted on top of a microscope slide coated with a film of petrolatum, and then 0.02g of the different oils or oil blends were spread over the top of the Vitro Skin. Water loss measurements were taken over time (1 data point/second) using a Courage and Khazaka TEWLmeter.[4]

Results and Discussion

Shower gels and body washes are cleansing formulations containing a blend of synthetic surfactants such as sodium laureth sulfate, cocoamidopropyl betaine and others in a product form that has become increasingly popular over the last decade. Isotropic shower gels contain 10-20% surfactants along with other minor ingredients such as fragrance, color, preservatives, emotive ingredients and/or particles for sensory or visual effects. Body washes are differentiated from shower gels by the high oil levels employed, mainly to support stronger deposition and moisturization claims. All of these formulations require a certain viscosity profile that makes them easy to pour from the bottle, but also easy to use and apply in the shower. The viscosity of simple surfactant blends in shower gels can be modified using salt (NaCl) which alters the surfactant physical chemistry or with various hydrophilic polymers such as polyacrylates which can be used to both thicken and/or suspend droplets or particles in the product.[5] Body washes use a surfactant structuring mechanism or different polymers to provide the high viscosity, lotion-like product properties that consumers prefer from these body cleansers.

Water-soluble silicones for modified lather and skin feel. Dimethicone polyether copolyols such as Silsoft* 895 and Silsoft 870 can be added to liquid body cleansers to modify lather properties to provide rich and creamy lather. These products are water soluble and they are typically added to shower gel formulations at 0.5-2% to modify lather during product use. Several silicone polyethers that can be used the for fine tuning of foam & lather and other specific characteristics of personal wash products are depicted in Figures 2 & 3. Also, silicone amines and quats such as Silsoft Tone, Silsoft A-553 and SME253 can also be added at similar levels to modify lather properties and in-use sensory performance as well as provide post-wash sensory benefits through deposition onto the skin during the shower. Their structures are depicted in Figure 4.

Since these modified silicones are water soluble and most likely are found as part of the surfactant system, they can impact the salt thickening capability of isotropic shower gel formulations. In these cases, polymeric viscosity modifiers such as carbomers are required to formulate a user acceptable final product.

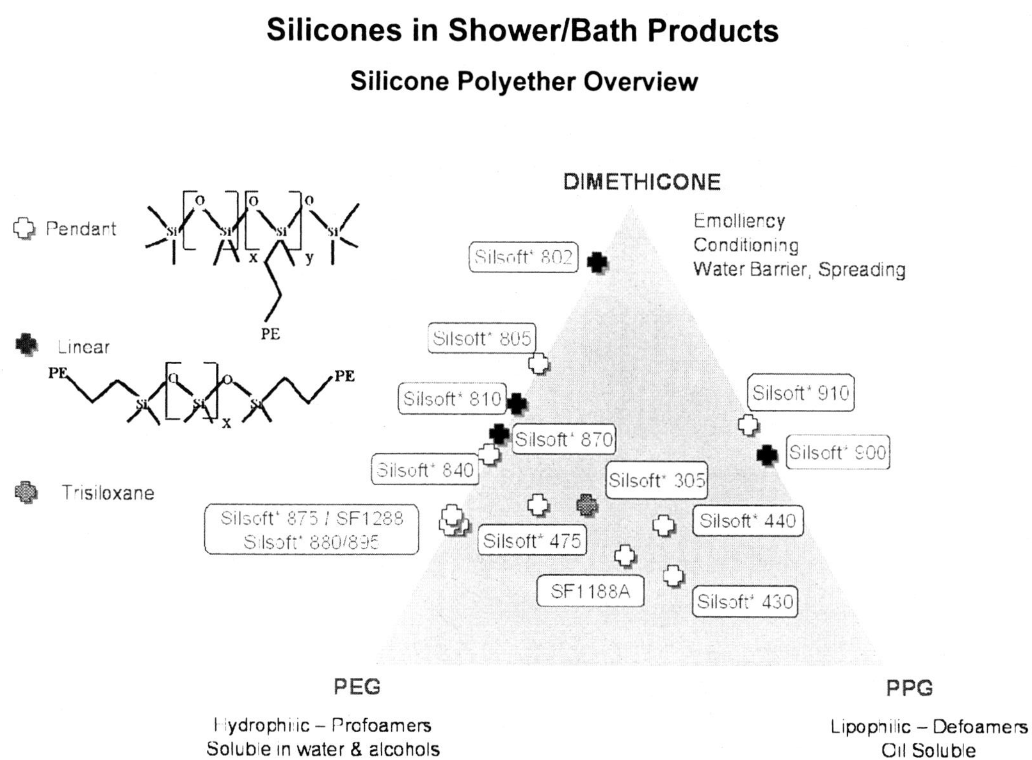

Figure 2: Silicone polyethers: Correlation architecture - functional benefits. PE = Polyether (PEG and/or PPG)

Silsoft* Copolyols as Foamants

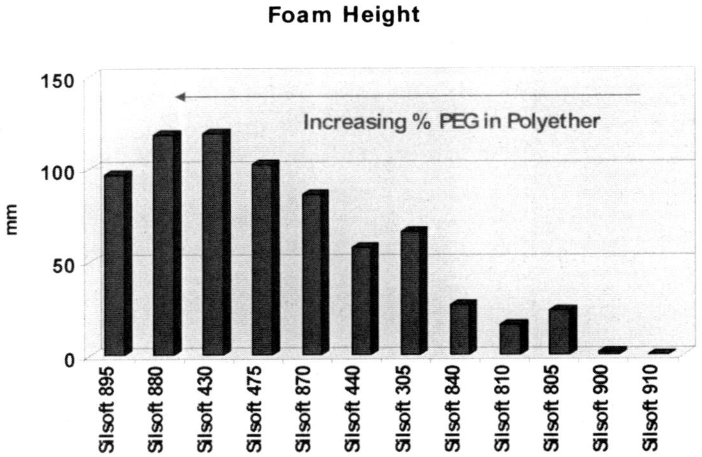

Foam increases as PEG content increases

Ross-Miles Foam Test [6] – 1% Silicone Polyether in Water (25°C)

Figure 3: Foam height [6] as a function of PEG content

Silicones in Shower/Bath Products

Silsoft* Tone (Polysilicone-18 Cetyl Phosphate): Linear silicone aminopolyalkyleneoxide copolymer containing silicone and polyether blocks neutralized with cetyl phosphate

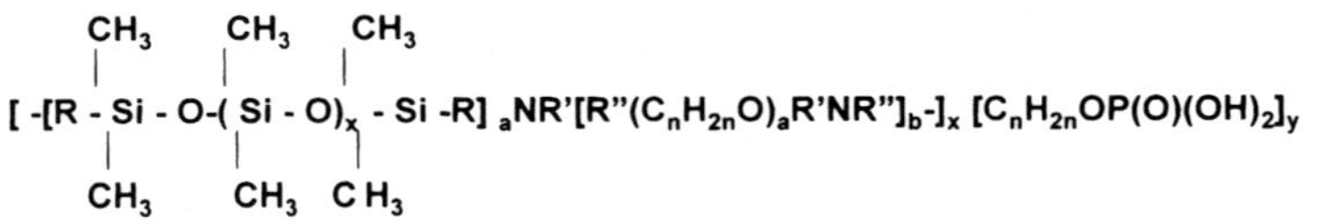

SME253 (Amodimethicone Microemulsion)

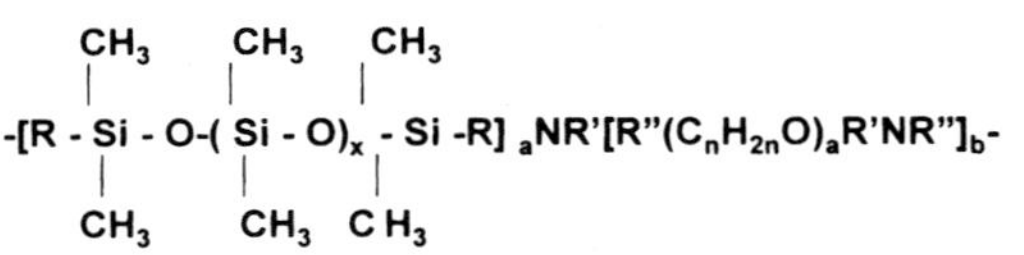

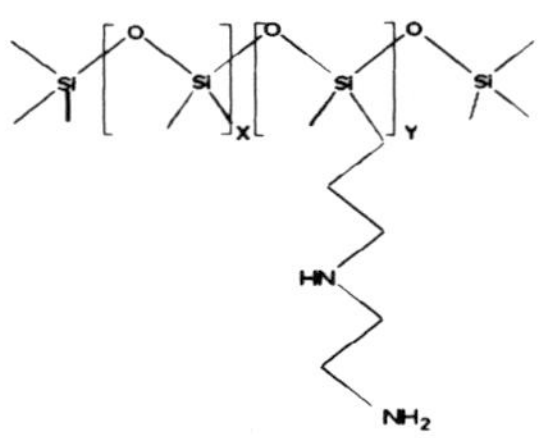

Silsoft* A-553 (Polysilicone-18): Block-copolymer of aminosilicone and polyalkyleneoxide

Figure 4: Structural depictions of amino silicones and quats

Silicones to modify the in-use and post-use sensory properties of moisturizing body washes. Moisturizing body washes typically contain high levels of petrolatum, triglyceride oils or other organic ingredients which are incorporated into the product at high levels to assure some deposition of the oil onto the skin. Various high viscosity silicone fluids (eg. SF96-1000) can also be formulated into moisturizing body washes to provide a differentiating in-use and/or post-use sensory benefits. Also, modified silicone fluids such as Silsoft*034, SF1642 and SF1632 can be blended with organic oils or high viscosity silicone oils to improve the feel of those ingredients when deposited on skin and still provide important skin moisturization benefits.

Silicone polymers can also be effective in providing enhanced fragrance deposition and longevity in traditional shower cleansing formulations, something that has always been difficult to achieve. One way to improve the fragrance properties of these products is to use the high levels of oils to carry fragrance to the skin. To achieve this goal, it is critical to ensure that the fragrance is solubilized by the oil phase and not just the surfactant phase. If the fragrance does partition into the oil phase, the oil can act as a carrier for the fragrance as it is deposited on the skin during the wash.

When a blend of high viscosity silicone oils, such as SF 96-1000 and alkyl modified silicones such as SF 1642 are formulated in a standard body wash, the silicone blend is deposited on the skin during the wash. This blend of SF96-1000 and SF1642 (4:1) is also an excellent carrier for fragrances to enhance fragrance deposition and longevity after wash as highlighted in Figures 5 & 6.

Data in the graphs presented in Figures 5 and 6 were obtained using SPME (solid phase microextraction) over samples of the body wash product itself or

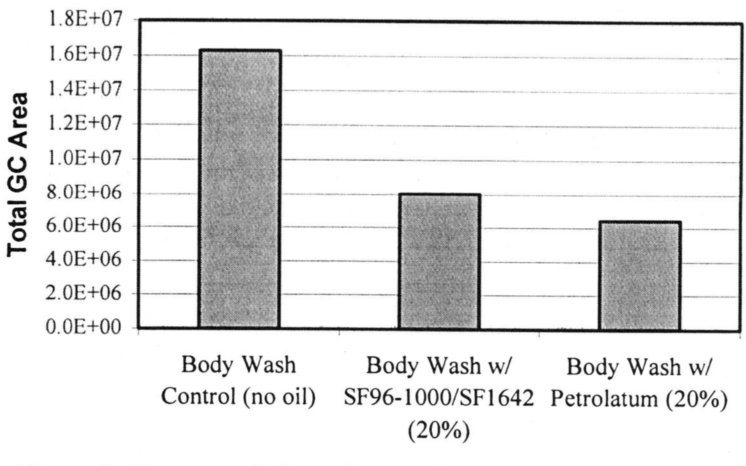

Figure 5. Fragrance Release from Body Washes (SPME/GC-MS)

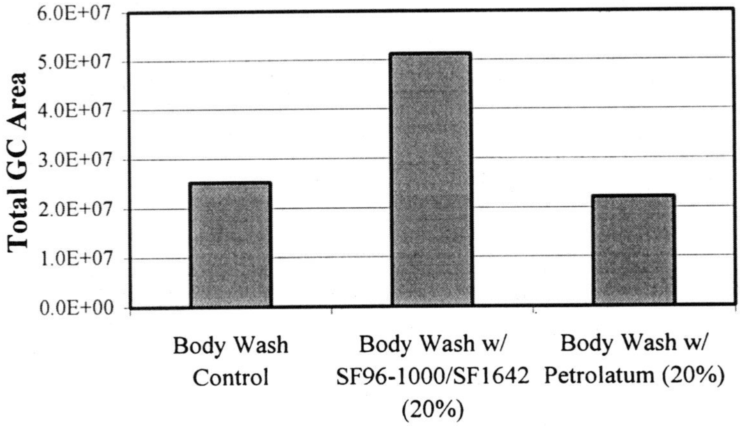

Figure 6. Fragrance Deposition on Skin from Body Washes (SPME/GC-MS)

over arms that had been washed (rinsed and dried) with the different body wash products. In this case, the body wash formulations contained 20% of the silicone blend (SF 96-1000/SF 1642 (4:1)) or 20% of petrolatum. Both oil matrices solubilized the fragrance resulting in less fragrance in the headspace over the product (Figure 5). These results indicate that some of the fragrance has partitioned into the oil phase, and therefore if the oil is deposited on the skin, it will carry the fragrance with it. The results in Figure 6 show that the amount of fragrance released from the skin after wash was much higher with the blend SF96-1000/SF1642 (4:1) than with the petrolatum indicating that the silicone oil blend is a more effective fragrance carrier than petrolatum in this application. These fragrance measurements should correlate directly with the fragrance impact the consumer actually perceives on their skin after using the product. Figures 7 and 8 show the chemistries behind Silsoft 034, SF-96-1000, SF1632 and SF1642.

Silicones for shower conditioners. A new offering in the shower product market are shower conditioners or in-shower body lotions. These products are to be used after cleansing in the shower to provide moisturization and fragrance benefits. Similar in concept to hair conditioners, these products are formulated with different oils and emotive ingredients, with typically only enough surfactant to emulsify the oil. Therefore these systems are quite high depositing formulations and the delivery of reasonably high levels of moisturizing ingredients to skin in the shower is possible.

While moisturization benefits are the basic claims for these products, good aesthetics both on wet and dry skin are critical for consumer acceptance. Some

Silicones in Shower/Bath Products

Silsoft* 034:
Caprylyl Methicone

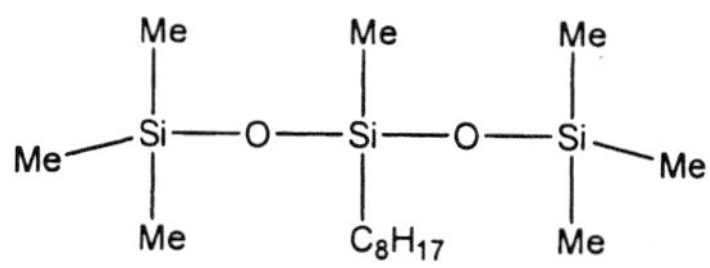

- **Product Attributes/Performance Benefits**

- Low Viscosity Oil
- Semi-Volatile
- Compatible With Cosmetic Oils
- Lowers Surface Tension of Oils
- Improves Spreading
- Emolliency
- Silky Soft Feel

*Figure 7: Structural depiction of Silsoft*034*

Silicones in Shower/Bath Products

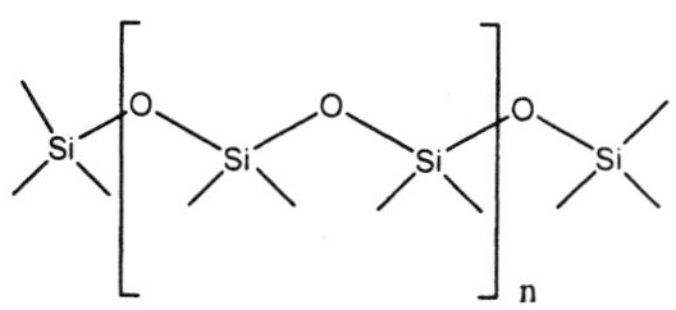

SF96-1000

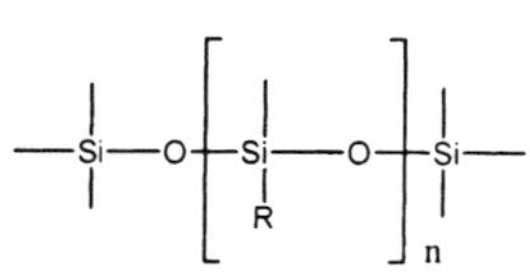

SF1632; R = Cetearyl

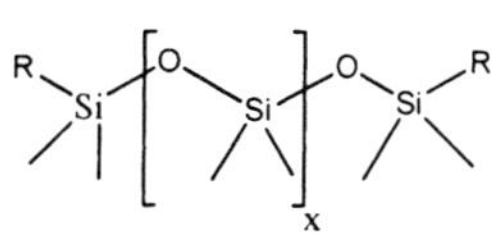

SF1642; R = C30-45

Figure 8: Structural depictions of SF96-1000, SF1632 and SF1642 specialty silicones from GE can provide unique sensory benefits both during and after use of the shower conditioner. Therefore, sensory panel evaluations were conducted to determine the impact of a typical silicone gel and modified silicones either by themselves or blended with organic oils. The panelists were asked to rate sensory attributes during application on the forearm on a scale of 0-10. The following graph (Figure 9) summarizes the results of 12 naïve panelists.

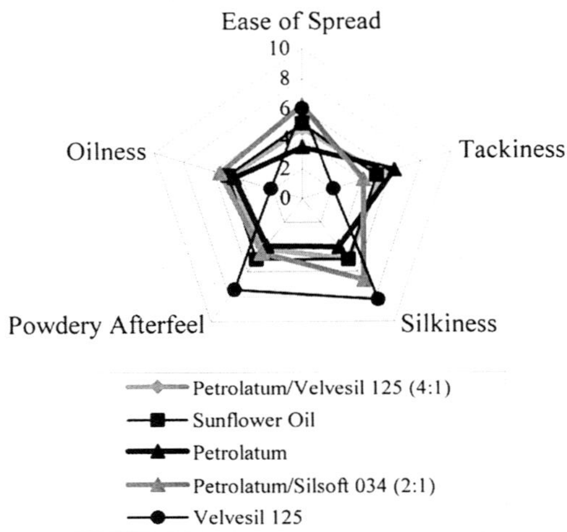

Figure 9. Naïve Sensory Panel Results for Oils and Oil Blends

The sensory panel data suggests that blending a silicone gel such as the patented Velvesil*125 [7,8] and/or Silsoft*034 with petrolatum, greatly diminishes the tackiness of the oil and provides a unique silky skin feel. Silsoft 034 is a trisiloxane that can lower the surface tension of organic oils and thereby promote the spreading of these materials on the skin, which has been reported to have beneficial properties for example in sun care.[9]

Silicone gels are very different from organic oils in that they are very silky and powdery without the tack and oily characteristics of organic oils. Specifically here, Velvesil 125 is a GE proprietary silicone 3-dimensional-copolymer network swollen in cyclopentasiloxane, providing a soft, powdery and somewhat dry sensory experience, not unlike a mild "squeaky clean" perception. It is a slightly hazy product with high diffuse transmittance that is capable of blurring fine lines and wrinkles. Also, it carries a C>30 alkyl functionality, which renders it more compatible with organic oils, unlike many other conventional silicone gels. Velvesil DM is a similar gel with a shorter C<20 alkyl function, swollen in 5 cSt polydimethylsiloxane (PDMS). It is less hazy, and provides a very smooth and almost therapeutic sensory experience, without exhibiting any negative perception of tackiness or oiliness. Both gels can easily be formulated into finished products and can be combined with other silicones, and are schematically depicted in Figure 10:

Silicones in Shower/ Bath Products

3D-Silicone Gels for Superior Sensory

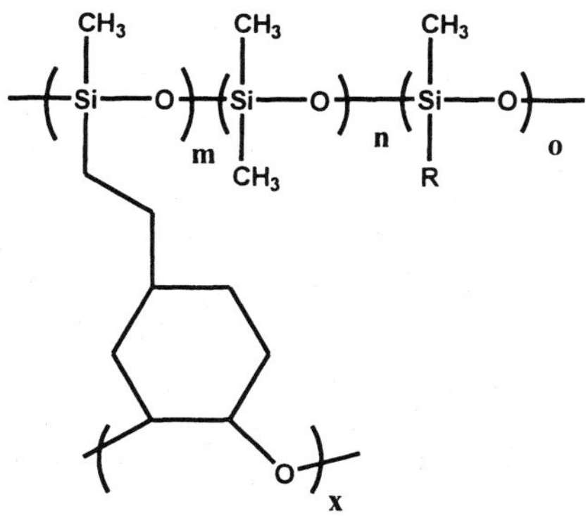

Velvesil* 125: R=C_{30-45}: Cyclopentasiloxane (and) C30-45 Alkyl Cetearyl Dimethicone Crosspolymer

Velvesil* DM: R=C_{14-20}; 5 cSt PDMS: Dimethicone (and) Cetearyl Dimethicone Crosspolymer

Figure 10: Structural depictions of Velvesil 125 silicone copolymer network and Velvesil DM siliconecrosspolymer network

Indeed, it has been found that when Velvesil 125 and Velvesil DM or Silsoft*034/petrolatum blends are incorporated into a fully formulated shower conditioner, the products provide a unique feel on both wet and dry skin. The aesthetics are pleasing and the wet skin does not feel tacky and oily like conditioners formulated with petrolatum or other organic oils.

While the sensory properties of these products are definitely key to consumer acceptance, moisturization claims are also important, so that occlusives such as petrolatum must maintain this behavior when blended with different silicone ingredients. Using Vitro-skin™ as a substrate, trans epidermal water loss (TEWL) measurements over time were made by measuring weight loss from the skin and by using a TEWLmeter.[10] In Figure 11, blank # 1 refers to untreated vitro skin, yielding TEWL values between 40 and 50 g/hm^2. When Velvesil DM was applied, the water loss is somewhat diminished, and lies somewhere between 33 and 40 g/hm^2. Petrolatum itself shows the lowest TEWL values with about 5 g/hm^2, and when mixed with Silsoft 034 or Velvesil 125, the respective TEWL data increase only slightly to between around 8 (Velvesil*DM) and around 10 (Silsoft*034) g/hm^2. It becomes thus clear, that when put in perspective, this small increase in TEWL when having petrolatum-silicone mixtures over petrolatum alone are insignificant when compared to the TEWL values of the untreated substrate, even in the case where 33% of silicone is employed. Further studies of these oil blends using microscopy confirmed that there was still crystalline behavior of the petrolatum blend indicating that the silicone ingredients did not dissolve the higher molecular weight waxy portion of the petrolatum and hence most likely explains why the occlusive nature of this material is maintained.

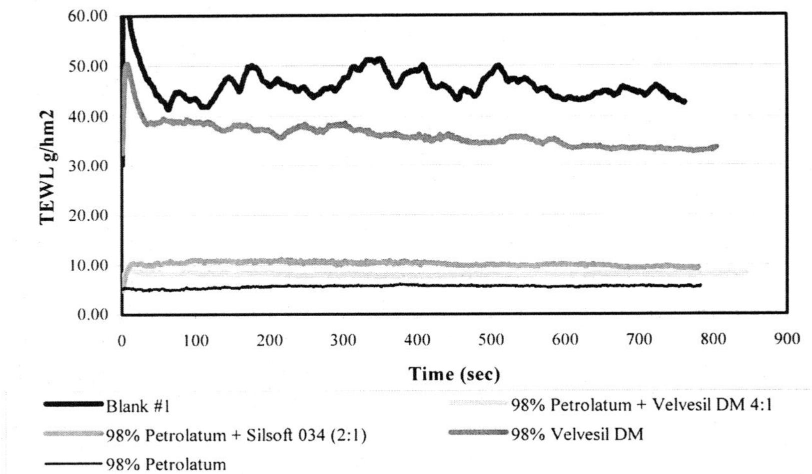

Figure 11. Skin water loss measurements using a TEWLmeter

Conclusions

In summary, specialty silicone fluids and modified silicones available from GE offer numerous benefits when formulated into different shower products. They can modify the lather and vary the in-use and post-wash skin feel sensory enhancement of cleansing products. Silicone blends can provide enhanced fragrance deposition and longevity from cleansing products and can offer sensory modification of different organic oils. In general, product sensory attributes can be "dialed into" shower/bath products through careful selection of the type and level of the silicone and the overall composition of the base shower/bath formulation.

Acknowledgements

The authors would like to acknowledge the contributions of colleagues at GE Advanced Materials, Silicones, to this publication: Janet Duffy, Lara Fieschi-Corso, Jerry Murphy, Anna Czech and Beatriz Blanco.

References

1. MacDonald, V. *Household and Personal Products Industry (HAPPI)*, **2004**, 11, 65-74
2. Mookherjee, B.D; Patel, S.M.; Trenkle, R.W.; Wilson, R.A. *Cosmetics and Toiletries*, **1998**, *113,* 53
3. Mookherjee, B.D.; Patel, S.M.; Trenkle, R.W.; Wilson, R.A. *Perfumer and Flavorist*, **1998**, *23*, 1.
4. TEWL: trans epidermal water loss
5. Lochhead, R.Y.; Fron, W.R. *Cosmetics and Toiletries*, **1993**, *108,* 95.
6. ASTM D1173-53; Gohlke, F.J. Parfuemerie und Kosmetik, **1964**, *45*, 59.
7. Thompson, A.; O'Brien, M.J. *Soap & Cosmetics*, November **2002**
8. US 6,444,745; US 6,531,540; US 6,538,061
9. Rojas Wahl, R.U.; Nicholson, J. R. *Seifen Oele Fette Wachse (SOFW), English Edition*, **2005**, 131(11), 42-52
10. Fromder, A.; Lippold, B.C. *International Journal of Cosmetic Science*, **1993**, *15*, 113.

Commercial and Regulatory Considerations for Cosmetic Nanotechnology

Chapter 11

Nanotechnology and Applications in Cosmetics: General Overview

Linda M. Katz

Office of Cosmetics and Colors, Center for Food Safety and Applied Nutrition, Food and Drug Administration, 5100 Paint Branch Parkway, HFS–100, College Park, MD 20740

Nanoparticles are defined as particles in the atomic, molecular or macromolecular levels ranging in size from 1-100 nanometers (nm); although for some novel properties the definition may be altered to take into account particles whose size may be <1m, or > 100 nm. As the commercial applications of nanotechnology have increased over the past several years, there has also been an increase in the potential uses of nanoparticles in cosmetic products as well as other products regulated by the Food and Drug Administration (FDA). There is a concern that substances considered safe for use as microparticles may penetrate more readily through skin as nanoparticles and exhibit different physical and chemical properties. The use of nanotechnology in cosmetics is not new, but dates back to 1961 with the advent of liposome technology use to market some moisturizing creams. This technology, which was used to alter optical properties, increase solubility and alter physical properties, provided for hydrophilic vesicles with phosphatidylcholine membrane(s) ranging in size from 15-3500 nm. The presentation below will address some of the current uses of nanotechnology in cosmetics marketed in the United States and the regulatory implications, as well as describe an FDA collaborative project with other Federal agencies and academia.

The Regulatory Structure

The mission of the Food and Drug Administration (FDA) is to ensure that drugs, medical devices, and vaccines are safe, effective, and properly labeled, and that food and cosmetic products are safe and properly labeled. The FDA paradigm for regulation of these products is based on a "risk management" framework that is applied on a product by product basis. However, some of the products that the FDA regulates, such as cosmetics and foods, do not require prior approval by FDA. The regulations stipulate that manufacturers of these products must ensure that they are safe prior to marketing.

Cosmetics cover a wide variety of products including lipsticks, fragrances, hair dyes, body lotions, shampoos, and are defined in the Federal Food, Drug and Cosmetic Act FD&C Act) as (1) articles intended to be rubbed, poured, and sprinkled on, introduced into, or otherwise applied to the human body or any part thereof for cleansing, beautifying, promoting attractiveness, or altering the appearance, and (2) articles intended for use as a component of any such articles; except that such term shall not include soap. (1) Additional information on the regulation of cosmetics may be found at http://www.cfsan.fda.gov/~dms/cos-toc.html. The distinction between a cosmetic and a drug product may be confusing in some cases. Drugs are defined by their intended use in the FD&C Act as "(A) articles intended for use in the diagnosis, cure, mitigation, treatment, or prevention of disease... and (b) articles (other than food) intended to affect the structure or any function of the body of man or other animals.(Ref. 201(g)(1)). The intended use of a product is determined, for example, by evaluation of labeling claims, advertising matter, or oral or written statements by those (or their representatives)legally responsible for labeling. While sunscreens are regulated as drugs in the U.S. such a product containing a moisturizer is also a cosmetic. Similarly, toothpastes that contain fluoride or deodorants that are also antiperspirants are regulated as both drugs and cosmetics. Importantly for our discussion, the intended use of a product will determine its regulatory status, not the use of nanoparticles in the product's formulation.

Nanotechnology – Introduction

FDA defines nanotechnology as the existence of materials or products at the atomic, molecular, or macromolecular levels, where at least one dimension affects the functional behavior of the drug/device and is in the length scale range of 1-100 nm. (2). To put this in perspective, 1 nm is equal to 10 hydrogen atoms laid side by side. The term "smart materials" (3) has been used to describe nanoparticles developed to target the delivery of drugs to specific organs of the body. Nanomaterials have many possible applications, including potential uses in cosmetic products, both in dry powders or liquid form. (4) The potential benefits afforded these products include having the property of not scattering visible light but still having sufficient size to scatter UV light (5). These properties make nanoparticles of sunscreen ingredients (i.e. titanium dioxide) transparent to the eye but still able to block the absorption of UV light (5).

Nanoparticles

Currently nanotechnology has been used in nanodispersed systems, nanocapsules, polymer systems and metal oxide nanoparticles. There are a variety of nanodispersed systems: liposomes, nanoemulsions, lipid nanoparticles; the solid core can be hydrophilic or hydrophobic depending upon the number of phospholid layers and deliver water or fat soluble actives. (Figure 1a, 1b, 1c, below) Nanoemulsions have been useful in typical cosmetic formulations, especially those found in skin and hair care preparations, because the light or oxygen sensitive ingredients can be protected and the emulsions have low biotoxicity.(6)

Nanotechnology has been used in cosmetics since 1961 with the advent of liposome technology used to market some moisturizing creams. This technology, which was used to alter optical properties, increase solubility, and alter physical properties, provided for hydrophilic vesicles with phosphatidylcholine membrane (s) ranging in size from 15-3500 nm. (7) The additional advantage was no or low carrier biotixicity. (8) Most of the nanodispersions used in cosmetics still are lipophilic vesicles ranging in size from 50-200 nm. Products having nanoparticles in the range of 50-100 nm have been used to produce translucent lotions. In 2001, between $25-30 million was attributed to the U.S. consumption of particulate delivery systems, of which 30% was from nanoscale material. (9)

Solid hydrophobic nanospheres, which have a high cationic charge density of the surface layer, tend to have strong interaction with skin/hair. The solid hydrophobic core protects water soluble and volatile ingredients. Examples of this use are found in "nutraceuticals", fragrances, and vitamins. Polymer nanoparticles can lead to a burst or controlled release of active ingredients and are considered to be to be more robust compared to liposomal formulations. They are found in a great number of naturally occurring and synthetic sources and are stable in both liquid and powder form.

Metal Oxide nanoparticles

Of particular interest are the metal oxide nanoparticles, such as titanium dioxide (TiO_2) and zinc oxide (ZnO), that are found in many commercial

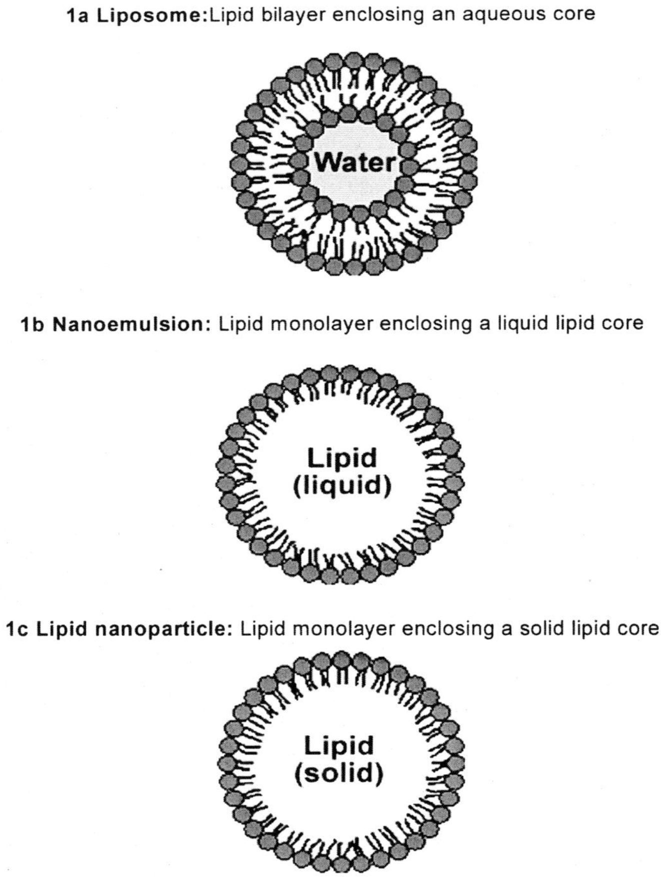

Figure 1. Structure of nanodispersed systems.
(Reproduced with permission from HAPPI *March 2002. Copyright 2002.)*

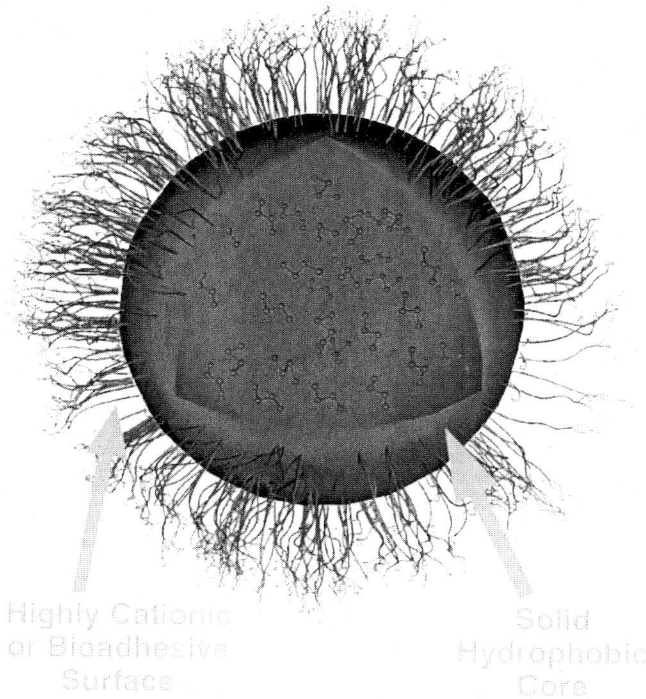

Figure 2. Schematic Image of Nanosol

applications, including skin lotions and sunscreens because they alter UV absorption properties and appear to be clear or translucent on the skin. (5) Uncoated TiO_2 absorbs photons of light and emits an excited electron. The electron can be transferred to O_2 and absorbed into dermal layers; irradiation by sunlight can lead to oxidative damage. (10) Coating stops the formation of reactive species and prevents agglomeration of particles. The small crystal size and controlled particle size give excellent dispersibility, attractive skin feel and a transparency on the skin. (11)

Current State of Activities

The potential for toxicity from the use of nanoparticles has attracted special attention; for example, increasing the hydroxyl substitution decreases toxicity of buckyballs (*12*). Nanotubes which have potential application in aerospace, fuel cell, computer and electronic industries have been shown to cause pulmonary toxicity and have a potential inhalation threat. (*13*) Prolonged exposure to airborne carbon nanotubes in mouse studies has shown lesions and inflammation of the lung. (*13*) Federal initiatives are underway to address the new safety concerns that arise. Importantly for cosmetic applications, the extent to which nanoparticles can be absorbed into the skin is not yet clear.

The Interagency Working Group on Nanotechnology Environmental and Health Implications (NEHI) was established by the Nanoscale Science, Engineering and Technology (NSET) subcommittee of the National Science and Technology Council (NSCT). The working group is comprised of individuals from a variety of agencies who have been tasked with looking at policies and issues that span the government and developing solutions to overcome these challenges. Federal agencies represented in NEHI include FDA, EPA, USDA, NIOSH, OSHA, NIH and NSF.

FDA collaborative projects include: (1) project with NIST that was begun in 2004, with the objective of developing expertise in nanotechnology in order to aid in the review of applications from sponsors of products that employ nanoparticles and to assist in writing guidance and setting standards for assays measuring biological responses. (2) Projects with NTP: 1st study is designed to examine the effect of nanoparticle size on skin penetration and the 2nd study is to examine the ability of titanium dioxide and zinc oxide to penetrate excised human skin over 24 hours. In addition, the FDA has formed an InterCenter working group that meets quarterly to discuss the status of products formulated with nanoparticles, as well as regulatory issues and implications. Some of these issues include the regulation of combination products, regulation of categories of products over which FDA has no premarket approval; and where in the product development nanotechnology may occur.

At the present time, FDA is aware that several products under its jurisdiction employ nanotechnology, including a few cosmetic products that claim to contain nanoparticles to increase the stability or modify release of ingredients as well as the moisturizing creams, referred to above, that employ nanotechnology.

Where do we go from here? We anticipate continued development and research in the area of nanotechnology and its use with encapsulation as well as new delivery systems to protect unstable ingredients,. However, as this technology continues to blossom we need to pay careful attention to toxicities that may develop.

References

1. Federal Food, Drug, and Cosmetic Act, As Amended; Sec. 201, Washington, D.C. 1993; p. 4.
2. FDA Nanotechnology, www.fda.gov/nanotechnology/
3. Lobenberg, Raimar, Smart Materials: Applications of Nanotechnology in Drug Delivery and Drug Targeting, IEEE, 82-83, July 2003.
4. Nanocapsules; Holster, P, et al; Technology White Papers; Cientifica; 10, October, 2003.
5. Dagani, R, Putting the 'nano' into composites, C&E News, 77, 25-37, June 7, 1999.
6. Lautenschlager, H., Liposomes, in Handbook of Cosmetic Science and Technology (Eds. A.O. Barel, M. Payne and H.I. Maibach), Marcel Dekker, Inc, New York, 201-209, 2001.
7. Solid Lipid Nanoparticles: Production, Characterization and Applications; Mehnert, W, Mader, K; Advanced Drug Delivery Reviews; 47, 165-196, 2001.
8. New York Society of Cosmetic Chemists Tech Notes, www.nyscc.org/nems/archieve/tech0402.htm
9. Vitamin A-Loaded Solid Lipid Nanoparticles for Topical Use Drug Release Properties; Jenning V, Schafer-Korting M, Gohia S; Journal of Controlled Release, 66, 115-126, 2000.
10. Dunford, R., Salinaro, A, Cai, L., Serpene, N., Horikashi, S., Hidaka, H., and Knowland, J., Chemical oxidation and DNA damage catalysed by inorganic sunscreen ingredients, FEBS Letters, 418, 87-90, 1997.

11. Aromatics & More www.aromaticsandmore.com/products/vitamims_sunscreens.html
12. Oberdorster, E.,Manufactured nanomaterials (fullerenes, C60) induce oxidative stress in the brain of juvenile large-mouth bass, Environ. Health Perspect. 112. 1058-1062, 2004
13. Lam, C-W, James, J.T., McCluskey, R., and Hunter, R.L., Pulminary toxicity of single-wall carbon nanotubes in mice 7 and 90 days after intratracheal instillation, Tox. Sci, 77, 126-134, 2004

Chapter 12

Personal Care and Cosmetic Nanotechnology: Perspective and Opportunities

K. O. Havelka

Consumer Products, Lubrizol Corporation, 29400 Lakeland Boulevard, Wickliffe, OH 44092

An overview of the challenges and opportunities in the fast-paced and dynamic environment of personal care and cosmetic nanotechnology that exist for chemists and engineers today will be reviewed. Increased understanding of structure-property relationships has enabled development of materials with diverse properties. Highlights of some new materials and delivery mechanism used in skin care and hair care will be reviewed. This will be coupled with a discussion of the multidisciplinary approach required to design formulations that balance surface (skin and hair) compatibility and functionality with aesthetics and appearance to provide products that consumers want. This chapter will conclude by summarizing current technology needs and identifying some emerging opportunities for scientists in personal care and cosmetic nanotechnology to design products that offer a viable alternative to cosmetic surgery.

Personal care and cosmetic nanotechnology is one of the most diverse areas in the chemical industry. It represents a blending of art and science that enables development of products that provide high levels of efficacy and are aesthetically pleasing. Based on the diverse requirements for personal care products, there is considerable opportunity for collaboration within and extending beyond the scientific community. This area is truly global and is highly valued by consumers. In 2004, the global cosmetics and toiletries market was valued at $230 billion and continues to grow with sales distributed globally as shown in Figure 1 (1). Western Europe, Asia Pacific, and North America represent the largest current market whereas Eastern Europe and Latin America are the fastest growing markets (2).

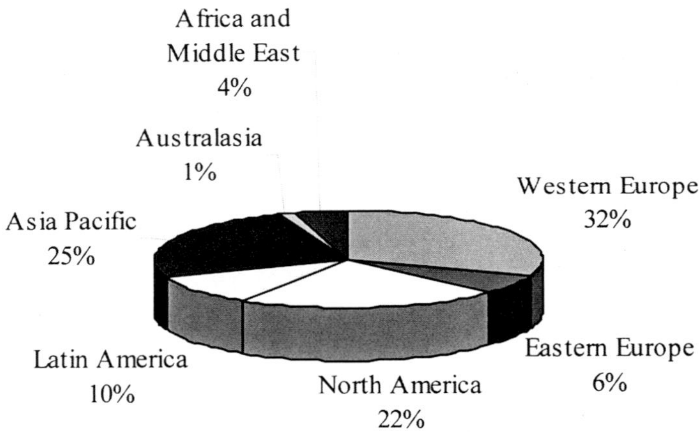

Figure 1. Global Cosmetic and Toiletries Market in 2004 $240 Billion

Although the market is global, consumer trends continue to exhibit a degree of regional specificity. This specificity is often influenced by skin and hair texture and pigmentation as well as regional preference. An example of a regional preference is the trend toward skin whitening products in the Asia-Pacific region. Many Japanese and Chinese consumers are looking to achieve fairer skin tones; this makes whitening products extremely fashionable in this region. Further segmentation can also be done by age and gender creating new product opportunities that are enabled by new chemicals.

Personal Care Technology Drivers

Personal care and cosmetic market segmentation by age, sex and ethnicity, is providing global opportunities for innovative technologies. About $7 billion of chemicals are sold to the cosmetics and toiletries market, with about half of these sales in specialty ingredients, including: surfactants, emollients, ultraviolet absorbers, and hair fixatives (3). Sales in this fast growing chemical area ranges from Baby Boomers with expendable cash looking to ward off aging to Gen-Xers craving an individual look; these diverse needs drive technology. The key technology drivers for the personal care and cosmetic market are listed below, including:

- Global expansion – Products to meet the growing demands for middle class in emerging regions.
- Increased market segmentation – Products that are focused on specific consumers such as by age, sex or ethnicity.
- Aging "boomers" – Boomers are at the height of their spending power and they are looking for products that provide a youthful look and even help ward off the effects of aging.
- Younger buyers - teenagers have access to an unprecedented amount of disposable income, accounting for up to 20 percent of total cosmetics and sales in the U.S. (3). Also teenagers' purchasing patterns are less established than older consumers providing manufacturers significant opportunities to introduce new products and mold future habits and brand loyalty. In addition, the highly fashion-conscious nature of younger consumers provides a further stimulus to manufacturers' innovation strategies, presenting a constant demand for new product.
- Natural ingredients – cosmetic active ingredients will continue to lead all other specialty chemicals used in personal care and cosmetics, such as, vitamins, botanical extracts, and polysaccharides (4).
- Increased efficacy demands – The cosmetics market is becoming very influenced by minor surgical cosmetic treatments and is beginning to offer consumers a credible but less expensive and less painful alternative to their surgical counterparts.
- Focus on wellness – The whole body approach is being taken in cosmetics, i.e., beauty from the outside in.
- Ingredient flow from "Prestige" to "Mass Market" – innovative ingredients are more rapidly making their way into all products, e.g., alpha-hydroxy for anti-aging.
- New forms and methods to deliver actives – nanotechnology to deliver incompatible actives and ultraviolet inhibitors.

Due to the diversity of problems to be solved, personal care and cosmetic nanotechnology requires a multidisciplinary team approach. This has attracted a broad spectrum of scientists to design formulations that balance compatibility and functionality with aesthetics and appearance to provide products that consumers want.

Considerations that Drive Consumer Purchases

Consumers wants appear almost limitless. However, there appears to be four considerations that drive purchasing of products, including: appearance on shelf, use of product, after feel of product and the long-term effects of the product. Each key driver will be explored in greater detail.

The appearance of product on shelf

The appearance of the product on the shelf includes its packaging, visual appeal, and the promises that it makes in the form of claims. Packaging is one of the key determinants for the consumer and for delivery of product. The package can simply hold the product or it can impact performance such as foaming for cleansers. The types of packaging used are diverse, including: bottles with different caps, pump dispensers, foam dispensers, sprays, aerosols, two phase products (shake and use), two compartment products (isolate materials by putting in separate compartments), and single use packaging, e.g., individual wipes. In addition, the containers range from opaque to clear. Containers that are clear enable marketers to show the product and potentially capitalize on its visual appeal.

The physical properties of a product impact its appearance on the shelf, its feel in use, and its final properties. Color, clarity and sometimes pearlescence are the physical properties that are most evident to consumers. Viscosity gives the product body. Often thickeners are used to suspend visible active ingredients such as fragrance beads or other encapsulated ingredients in personal care products. The form of product also has a significant influence. It can range from a solid, solution, suspension, dispersion, or emulsion. Emulsions vary from water oil-in-water (O/W), water-in-oil (W/O), multiple emulsions (W/O/W or O/W/O). These forms offer the consumer many choices and the scientist considerable flexibility when developing a product.

It can be argued that although the appearance of the product is important to catching the consumer's attention, the most important thing to the consumer is the promise that the product makes on the label in its claimed performance. The claims sell the product the first time, but the benefits ensure the customer comes back. In regards to claims, there is ambiguity in the cosmetics industry of how claims should be supported (5). Often the proof is determined by the individual company. For the most part, this self regulation appears to work well, but it appears that there is more that can be done to ensure that accurate claims are being made and that consumer expectations are being met. Oversight for claims comes from the FDA and the Cosmetic, Toiletry and Fragrance Association (CTFA), and others offer some guidelines, but their opinions vary on the use of data. To satisfy the regulatory authorities and the consumer, it is vital to use multiple sources of data to support product claims, such as consumer testing, technical, marketing, and lifestyle data to provide a complete data set that validates product performance.

Technological developments and a wider use of innovative ingredients enable one to generate support for specific product claims. This is evident in the skin care market where often beauty and therapeutic claims are being made. Some innovative skin care products now includes formulas which promise to diminish wrinkles, acne marks, blotches, scars, red blotches as well as skin aging. These clinically-inspired home treatments are positioned for continued expansion, and their growth is supported by increasing consumer interest in improving one's look and backed by further technological discoveries and progress (6).

Use of Product

Personal care products are designed to stimulate the senses. The goal is to create a delightful experience that consumers will remember and come back for. Conventional formulations stimulate the senses of sight, touch and smell. Newer skin care formulations are looking to even stimulate the senses of taste and sound. When a product is applied, it should apply evenly and not be greasy; it should make a person feel pampered and cared for. To enable these diverse attributes, product customization, personalization and multifunctional properties are required. Figure 2 presents the formulation properties that control these sensory perceptions.

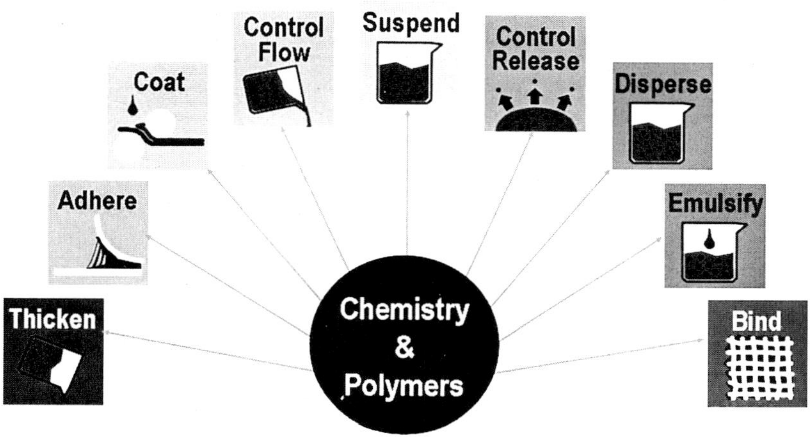

Figure 2. Diverse Properties in personal care products

In personal care formulations polymers dominate. Polymers are at the forefront due to their ability to perform multiple functions in a formulation ranging from film forming, to emulsion and dispersion stabilization, to rheology control. Hair care and skin care products require a delicate balance of efficacy and performance. Products that offer benefits for a variety of individual needs and styles including the ethnic population and specialty products for men are becoming an important trend (3). Cleaning of the hair and obtaining luster have historically been the focus of hair care products. Customers now want ingredients that deliver many attributes such as light or weightless conditioning, repair and conditioning, shine, style without residue, and strong hold. Some of the most desired hair care ingredients include conditioning agents, antifrizz agents, styling polymers, hair colorants, anti-aging actives, naturals, rheology modifiers, stabilizers, silicones and hair protectants to improve the appearance of hair and to hold the style. This has been done through both chemistry and processing.

Scientists have used clever processing and novel chemistry approaches to address the cosmetic market. Processing techniques such as high pressure and high shear emulsions have shown reduced irritation potential in sun screens (2). The types of emulsions available today are quite broad. They range from traditional oil-in-water to invert emulsion (water-in-oil) to 2-in-1 systems ranging from macro to micro-emulsions to clear emulsion. Clear emulsions contain chemicals with the same index of refraction for both the internal and external phases to maintain clarity. Cold processing has also become important. The ability to cold process has increased the possibility to work with heat sensitive materials to ensure optimum efficacy in cosmetic products. The types of chemistry and processing conditions used can have a significant impact on the final properties of the product.

After feel and impact of product

For a consumer to repeat the purchase of a product, it must go on evenly, not be sticky or greasy, and leave him feeling that it is beneficial for his well-being. This wellness trend is evident at retail stores; many retailers have aligned their beauty lines with non-beauty products, positioning health products such as vitamins in close proximity to cosmetics. Even some manufacturers have begun supplying complete wellness systems, expanding beyond cosmetics and toiletries to "beauty" products that enhance one's health from the inside (6).

In skin and body care, key advances have been made in product efficacy and delivery mechanisms to enable significant product enhancements. The types of chemistry used in formulating range from synthetic to all natural ingredients. Chemistry used to deliver actives vary in size and form from microcapsules, microspheres, nanoparticles to polymer skin products. Active ingredients range in type from antioxidants to nutraceuticals and cosmaceuticals. There is a strong demand for cosmeceuticals that address aging. Cosmeceuticals are products that not only enhance ones appearance but also have a positive physiological effect at the cellular level (6). Today's formulations also include vitamins and other nutritional ingredients that are referred to as nutraceuticals; nutraceuticals are nutritional ingredients that are used for their beautification and nutritional benefits (7). Nutraceuticals are often used in personal care products to reduce the visible effects of aging, acne, and impart other benefits. Another benefit that is being explored in body care is natural and novel ingredients such as epidermal growth factor, poly-collagen peptides and nanoparticles of silica and soyaprotein, that promise to penetrate, lift and tighten sagging skin (2). These ingredients are intended to improve the appearance of the skin by toning and firming without exercise or surgery. They offer the hope of getting rid of cellulite and ideally preventing it. The explosion of exciting new concepts and new properties being explored creates a significant opportunity for chemists to develop effective materials that improve appearance and provide long term health benefits.

Long term after effects

Consumers have come to expect not only primary benefits from personal care and cosmetic products, like cleansing and moisturizing, but also long term therapeutic effects like wrinkle reduction. In addition to increased performance expectations, consumers are significantly more aware of the functions of specific ingredients. This wellness trend of personal care products has driven increased investment in the research and development of formulations to contain natural and novel ingredients that impart a benefit. Increased demand for natural ingredients has lead to more formulations containing botanicals and vitamins and even a repositioning of existing products that contain naturally-derived ingredients like oleochemicals, natural gums, or starches (6).

In addition to natural ingredients, the increased demand for technologically advanced products has spurred the development of cosmeceuticals, products containing active ingredients that claim to provide a clinical effect in addition to their traditional cosmetic function. Consumers are most voracious for technical products that claim to offer some protection against the visible signs of aging, and manufacturers are obliging. A hot area of research in personal care products is to develop ones that compete directly with other measures such as minor surgery, injections and peels. A range of cosmeceutical ingredients to address these more challenging properties can be found in personal care products today, including coenzyme Q10, which is produced by the body and works as an antioxidant (4). The development of cosmeceuticals has spurred an explosion of new formulation concepts and new ingredients.

Sunscreen chemistry has been a key benefactor of the technology advances due to the desire to mitigate the damaging effects of overexposure to the sun. Photobiology and the emulsion technology of high SPF water-resistant formulations are an area of intense investigation to address this need. Active ingredients that protect the skin from the suns damaging rays tend to be a mix of organic and inorganics to provide UV-A and UV-B protection. Nanoparticles have recently received attention as potential active ingredients that would facilitate improved sun protection without the tradeoff of whitening seen with micron-sized TiO_2 on the skin (6). In addition, nanotechnology is being investigated as a means to compatibilize and potentially deliver active chemistry that otherwise would not be readily absorbed into the skin such as Vitamin E. Applications of cosmetic nanotechnology are both exciting and controversial and will continue to be an area of intense investigation by multidisciplinary teams.

Delivering actives in cosmetically elegant formulations is an area that has made significant progress over the past 20 years. Actives enable long term benefits of personal care product. The actives most frequently used fall into four different categories, including:

- Herbal – herbs for skin and hair care
- Animal extracts – proteins and collagen
- Plant derived products
- Innovative chemicals that can be used as vehicles – alpha-hydroxy acids

The focus of research on delivering actives is on getting any active out of the vehicle and into the target zone and at a concentration that is effective. Almost anything below a certain size will penetrate the skin; the more lipophilic something is the more likely it is to penetrate. The amount that penetrates the skin depends upon the nature of the ingredient, the vehicle, other chemicals in the formulation, concentration of the ingredient, application and physiological factors. The skin barrier function may vary according to age, sex and ethnicity. The challenge is to develop products that do what they are supposed to do and that are safe to use. Delivery methods for actives range from conventional means, such as encapsulation and microspheres, to newer methods like nanotechnology. Nanotechnology is an emerging technology that offers significant promise in personal care and cosmetics; some areas currently under investigation include:

- Sunscreens – nanoparticle zinc oxide can act as a sun blockers to protect human skin in formulation that go on smooth and silky clear

- Anti-aging – nanoscale encapsultation of retinol improves formulation compatibility
- Lotus effect coating (5) – self cleaning effect that is being explored for hard surface cleaning and may also work in personal care cleansers.

The research in effective delivery of actives is one of the most important activities in the personal care industry. Sunscreen development is the area that is receiving the most attention as the awareness of the damaging effects of sun on the skin is coming to the fore.

Sunscreen Development Case Study

Most undesirable skin conditions can be traced back to the deleterious influence of sunlight. Based on this, considerable research is going into developing new very water resistant and sweat resistant sun screen formulations that are gentle to the skin. The research ranges from new sun screen actives to new formulation development. An investigation of the ability for an invert emulsion, water-in-oil (W/O) to provide a high level of sunscreen efficacy and still feel light to the consumer is presented below.

Invert emulsions by nature provide superior water resistance than oil-in-water emulsions due to the external phase, oil, being water insoluble. Also invert emulsions are more lipophilic which makes them more compatible with the skin and may allow for better skin penetration of the sunscreen thus providing a higher level of protection. To further explore this, water-in-oil sunscreen emulsions ((sun protection factor) SPF-15, 30 and SPF-45) were prepared with poly(isobutylenyl) succinic anhydride based emulsifiers. These formulations were non-irritating and demonstrated good thermal and shear stabilities as well as a high level of water resistance (9).

One key feature of the sunscreen formulations are that poly(isobutylenyl) succinic anhydride (PIBSA) based emulsifiers are active at very low treat rates, ranging from .5 to 2%. The emulsifiers performed in all base oils typically used in sunscreen formulations, including: silicone, silicone-hybrid, ester and mineral oil formulations. The formulations cut across a wide range of viscosities. This versatility enables formulations to be prepared that have a wide range of physical properties. Sunscreen subject testing on the SPF-30 formulations confirmed sun protection factor of SPF 30 and very water-resistant performance. There were no adverse or unusual effects of the sunscreen observed on the test subjects (9).

Experimental

It was felt that the unique structural features of PIBSA would provide excellent emulsion stability and application properties to a very water resistant sunscreen. As shown in the model of PIBSA versus oleic and stearic acids in Figure 3, PIBSA is higher molecular weight compared to these conventional emulsifiers, therefore PIBSA should provide greater steric hindrance between molecules. PIBSA is also bulkier due to branching so it takes up more volume per area and this should provide greater emulsion stability.

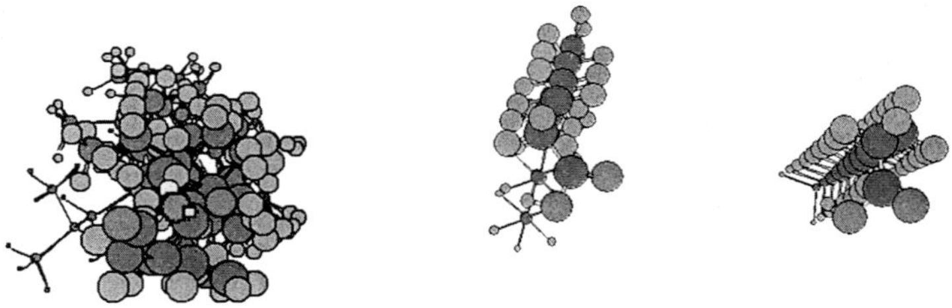

PIBSA Oleic Acid Stearic Acid

Figure 3. Structural Features of PIBSA versus Other Emulsifiers

In addition to PIBSA's size and shape, its succinic anhydride functionality provides an excellent reactive site for further functionalization with nucleophiles or hydrophiles. As shown in Figure 4, PIBSA can be functionalized by a variety of nucleophiles; polyols, polyamines and alkanolamines are the most useful for emulsion stability.

The emulsions prepared in this study showed excellent efficacy in sunscreen testing (10). Very water resistant sunscreen formulations were made and tested. Table 1 is a sample formulation that was made of the very water resistant SPF 30 sunscreen. This formulation was processed by heating both phases and then adding Phase B to Phase A to ensure good mixing and stability.

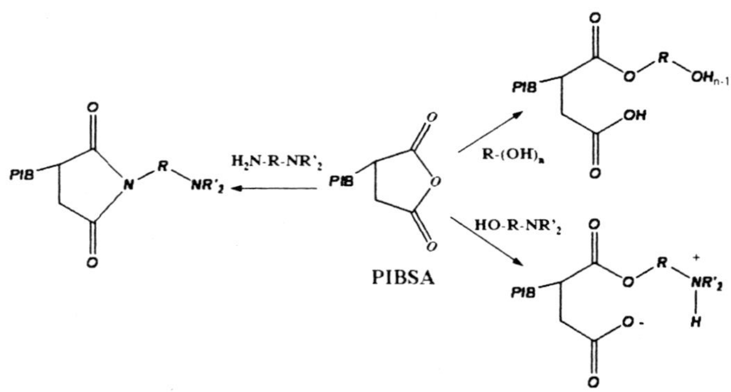

Figure 4. Functionalizing Poly(isobutenyl) succinic anhydride (PIBSA)

Table 1. Very Water Resistant SPF 30* W/O Sun Protection Cream

Ingredients:	**% Wt.**
Phase A	
PIBSA emulsifier	2.00
Dioctyl Sebacate	2.50
Uvinul MC-80	7.50
Eutanol G	7.00
Uvinul M-40	4.00
Siliconyl Beeswax	1.00
Castorwax MP70	1.00
Arlacel P-135	0.75
Cab-O-Sil M5	1.00
Z-Cote HP-1	7.00
Phase B	
Deionized Water	61.30
Plantaren 2000	0.10
Propylene Glycol	3.00
Magnesium Sulfate	0.75
Disodium EDTA	0.10
Phenonip	1.00
Total	100.00

Results and Discussion

Sunscreens based on the PIBSA emulsifiers proved to be very water resistant in in vivo very water resistant SPF testing on 3 subjects (10) as shown in Figure 5. The testing indicated that PIBSA emulsifiers performed comparably to functionalized silicone emulsifiers at a fraction of the cost. In addition, sensory testing indicated that the formulations were rated as cosmetically elegant. PIBSA emulsifier technology although new to personal care has been widely used in the automotive and industrial lubrication (11). Based on the excellent performance seen in the sunscreen applications, it appears that PIBSA technology can be more broadly applied to other personal care applications, e.g., skin lotions and creams, barrier creams, baby care, cosmetics, and ethnic hair care. This represents an area for further investigation by chemists and engineers.

Conclusion

It is clear that significant advances have been made in the chemistry, polymers, and delivery systems utilized in the personal care and cosmetic

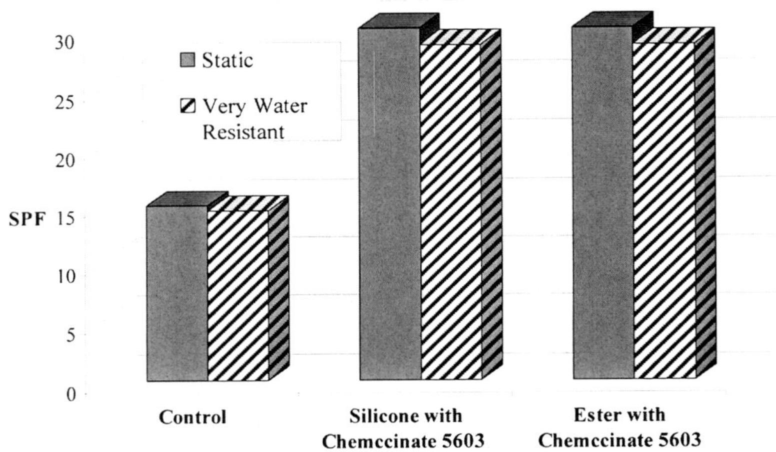

Figure 5. In Vivo Very Water Resistant SPF 30 Testing

nanotechnology. The focus on natural ingredients, innovative technology, nanotechnology, and delivery systems will continue to grow and have led to areas of research such as cosmeceuticals and nutraceuticals. It is conceivable that this may expand into other exciting new areas of research, including: neurocosmetics (emotional aspects of happiness); active cosmetics (reacting to changes in immediate environments); programmable cosmetics (changing with the environment for long-term needs); fast acting cosmetics (with clearly visible results in a short time); and nutritional cosmetics (internally providing cosmetic effects, something that is already happening) leading to new products with properties not seen today. These exciting potential developments among other are some of the things that scientists and engineers working together in multidisciplinary teams with physicians and other experts we may have to look forward to.

References

1. Euromonitor International, 2004
2. Van Arnum, P. *Chemical News and Intelligence*, May 2005.
3. D'Amico, E. *Chemical Week*, March 1, 2006.
4. Branna, T. *Household and Personal Products Industry*, March 2005.
5. *Soap, Perfumery & Cosmetics.* 78(1) p.16. January 2005.
6. de Guzman, D. *Chemical News and Intelligence*, January 2005.
7. Parker, S. *Soap, Perfumery & Cosmetics*. 78(6) p.32. June 2005

8. Zhiguang Guo, Feng Zhou, Jingcheng Hao, and Weimin *Liu Journal of the American Chemical Society*, **2005**; 127(45) pp 15670 – 15671
9. B. Filippini, J. Carey, S. Butz, N. Huang *Society of Cosmetic Chemists,* May 2003.
10. FDA Title 21, Subchapter D, Part 352 – Sunscreen Drug Products for Over the Counter, OTC, human use, Sec. 352.72-352.76
11. M.A. Mekewi *Materials Research Innovations*, October 2002, 6 (4) pp 214-217.

Indexes

Author Index

Subject Index

C

D

E

F

R

S

T

U

V

W

X

Z